Neurocultural Health and Wellbeing

Aim of this devoted book Series in neurology is to highlight the relationship between neuroscience and culture. Nowadays, there is more evidence of how our brain is influenced by the various artistic and cultural disciplines, both in the form of entertainment as well as – and above all – through the emotions and gratification which lead to wellbeing. The emotional mechanisms, the different cultural manifestations provided, are due to the activation of a perceptual and cognitive range that constitute the basis of the social behaviors. All this makes us aware of the benefits the arts have on personal and collective health.

This concept was already known to philosophers in ancient times, who were convinced that inner balance was influenced by culture and in particular by music.

During the centuries, men understood that the *cultivation of the spirit*, or *humanitas*, had a certain role on behavior and from the beginning of modern experimental science, in the fifteenth, the spread of the notions of neuroanatomy allowed artists to get closer to the knowledge of the brain's mechanisms and reveal the emotional and empathic responses at the basis of creativity, and indeed of their own psychophysical wellbeing. A dialogue between the science of the mind and artistic disciplines was born. The results of this meeting made it possible to better define which mental processes are involved when we come into contact with the various humanistic disciplines, and how they can be applied, for instance, to the treatment of mental disorders and neuropsychiatric diseases.

The goal of this Series is both to prove the role of biological brain mechanisms and the influence of various artistic forms on clinical practice, especially in neuropsychiatric disorders, as well as to trace different therapeutic and psycho-physical well-being applications based on scientific evidence from medical literature.

Volumes of the Series will be edited by experts under the supervision of an international editorial committee. Each book, focused on a specific discipline, will provide knowledge on relationship between the brain activity and different forms of language, communication, art. This close inter-relationship with specific focus on different forms of art will explain the effectiveness of this kind of approaches in neuro-psychiatric diseases.

This Series will allow to understand how the culture is one of the fundamental tools to improve general well-being, quality of life and motivation in neurological diseases.

This Series also find a correlation with SDG3 goal "ensuring healthy lives and promoting wellbeing for all at all ages" for the progress of the health and wellbeing considering that neurological diseases are on the rise worldwide including in the developing countries.

Lorenzo Lorusso •
Michele Augusto Riva •
Vittorio Alessandro Sironi
Editors

Effects of Opera Music from Brain to Body

A Matter of Wellbeing

Springer

Editors
Lorenzo Lorusso
Neurology Unit
A.S.S.T. Lecco-Merate
Merate, Italy

Vittorio Alessandro Sironi
History of Medicine
University of Milano-Bicocca
Milano, Italy

Michele Augusto Riva
School of Medicine and Surgery
University of Milano-Bicocca; Occupational Health Unit, Fondazione IRCCS San Gerardo dei Tintori
Monza, Italy

ISSN 2731-4464 ISSN 2731-4472 (electronic)
Neurocultural Health and Wellbeing
ISBN 978-3-031-34771-9 ISBN 978-3-031-34769-6 (eBook)
https://doi.org/10.1007/978-3-031-34769-6

This Springer imprint is published by the registered company Springer Nature Switzerland AG
The registered company address is: Gewerbestrasse 11, 6330 Cham, Switzerland

Paper in this product is recyclable.

Prologue

The very welcome proposal to write a short reflection on this precious book offered me the opportunity to reconnect the threads between my musical studies, carried out since childhood, and my subsequent university studies. During my years of study in the Faculty of Philosophy at the State University of Milan, a keen interest developed in researching the biological basis of moral behavior, leading to the definition of a specific branch called neuroethics. Paul Ricoeur and Jean-Pierre Changeux, in their *Nature and the Rule*, had already initiated a largely unsuccessful dialogue between science and philosophy concerning the investigation of the links between neurophysiological processes and our cultural architectures.

With the development of the recent studies presented in this volume, for the first time, we have a complete and comprehensive picture of the countless interactions between neuroscience and music. If a first glimpse was taken of the cognitive implications of music with Oliver Sacks and his well-known *Musicophilia*, here we can observe first-hand the therapeutic effects of opera in particular: from the role of "accompanied singing" in motor rehabilitation and the treatment of neurodegenerative diseases to the neural basis of musical imagination. The central role of audiovisual-motor mirror neurons in learning and listening to an opera offers a scientific explanation of the performance skills and the ability to coordinate between performers, thereby grasping the conductor's expressive intentions. Similarly, the neural mechanisms of the aesthetic musical experience shed light on how music is able to act on pathological areas related to depression, the emotional components of one's physiological state, the efficiency of cardiovascular activity, and much more. Hence, it is impossible not to recognize the central role of the opera experience for the individual as a whole. If our musical heritage constitutes a *unicum* in Western history, the research results proposed in this book truly open up the way to a new "philosophy of nature," which highlights a dense web of cross-references between different spheres: from the medical-scientific to music as the most complete expression of human faculties.

Milan, Italy | Michele Gamba

Ouverture

In recent years, music has become an object of interest to the medical world from multiple perspectives: neuroscientists study the relationship between music and brain functioning in the light of new findings in the fields of neurophysiology, neuroimaging, cognitive science, and neuro-musicology; clinicians investigate the therapeutic potential of music, especially in the field of treatment and rehabilitation of individuals with chronic-degenerative diseases; medical historians and medical anthropologists analyze the use of music in health care, in different historical eras and cultures. This interest in the relationship between music and medicine has led some scholars to investigate in greater detail the relationship between medicine and different genres of music, including opera.

Although the production of new operas declined significantly after the mid-twentieth century, partly due to the emergence of new forms of entertainment and spectacle, the number of fans of this musical genre remained significantly high, as did the number of performances given in theaters around the world: in the USA, there is estimated to be approximately one hundred opera companies in operation, and in the last decade, the Metropolitan Opera House in New York had an annual budget of more than $320 million. After the USA, Germany has the second-largest market for opera, followed by the other European nations—Italy, France, Austria, Norway, the Netherlands, Spain, Russia, and Hungary—and very promising markets include Australia, South Africa, and Argentina. Although the COVID-19 pandemic has hit this entertainment industry hard over the past two years, theaters are now starting to fill up again.

The prolonged and lasting success of some of the operas composed in past centuries derives from a combination of factors at the root of which, in addition to the quality of the music, is the dramaturgical effectiveness of the libretto and all the elements of which the performance is composed: in fact, this particular theatrical and musical genre requires that music and singing be combined with action on the stage. Those who attend an opera inside a theater or opera house hear not only the sound of the instruments emanating from the orchestra, but also the voices of the singers, with the many nuances and timbres that they carry. In addition, the spectator observes the colors and shapes of the sets prepared and the costumes worn by the performers while witnessing the staging curated by the director and the choreography performed by the dancers. For this reason, the relationship between opera and medicine appears to be of greater complexity to analyze, and this complexity prompted the editors of this volume to choose a multidisciplinary approach to its drafting. The authors are clinical physicians belonging to different disciplines (neurology, neurosurgery, and occupational medicine), historians, anthropologists, and musicologists, and each author has tried to give his or her point of view on the relationship between medicine and opera.

The volume has been divided into three sections, each called an "Act," using opera terminology. In Act 1, the theme of the representation of diseases and those who cure diseases within operas is addressed. Indeed, as is the case with other art forms, operas provide us with valuable information about the state of knowledge and perceptions regarding certain diseases, relative to the period of composition of the operas themselves, which offer important evidence about the beliefs of the time regarding diseases whose etiology was often still unknown at the time. Comparing operas from later periods can also allow us to observe changes in the epidemiological picture, with the disappearance of some diseases and the advent of new ones, as well as offer insights into how therapeutic approaches to health problems have changed over the centuries, and how the doctor–patient relationship has changed. The representation of healthcare workers in operas is another topic addressed in this Act 1: doctors and quacks, as well as herbalists and pharmacists, are frequently represented in operas, in both primary and secondary roles. Here again, it is possible to understand and analyze the perceptions and credibility that society of the time accorded to these figures. This first section also looks at the benefits that can be derived from listening to opera, in what is now called "opera therapy," recognizing it as a specific form of music therapy.

Act 2 of the volume is entirely devoted to the relationship between music and the brain. In operas, mental illnesses and neurological disorders have always been frequently portrayed, probably because they are more likely to attract the interest of the audience; as with other types of illnesses, the portrayal of neuropsychiatric illnesses can provide valuable information about the state of knowledge of these diseases in different historical periods. In a section devoted to neuromusicology, the effects of opera on the human brain were analyzed, and their possible application in the treatment of various neurological and psychiatric problems: the therapeutic potential is very high, although so far studies in this field are still scarce and limited to the treatment of dementia. Opera can bring about changes not only in the brains of those who listen to it but also in those of professional singers: although further neuroimaging and neurophysiology studies on larger samples of subjects are needed, some effects on neuroplasticity, derived from the vocal training of professional singers in operas, have already been shown.

Act 3 is devoted to descriptions of diseases that affect workers in the opera world. As was the case in the past, currently, the illnesses that most affect professional musicians and singers are hearing disorders, voice impairments, and musculoskeletal disorders, genuine occupational diseases that can be eligible for compensation. In addition to these physical problems, neuropsychological issues such as music performance anxiety, which can affect up to 25 percent of professional performers, should not be underestimated. This type of neuropsychiatric disorder was also very much present in the great composers of the past, and the biographies of some of them often echo the "genius and madness" cliché. In some cases, the damage suffered was a real obstacle to the creation of their work, while in other instances, the pathology supported the authors' creativity. References to occupational diseases are also present in the operas themselves. Beginning in the nineteenth century, society began to appreciate the damage caused by the most deleterious aspects of the Industrial Revolution to the health of the men and women who constituted the new proletariat, thereby also influencing composers and librettists, who included in their works some reflections on the meaning of work and its harmful effects on workers' health. Finally, the last part of Act 3 discusses the impact of the recent COVID-19 pandemic on the internal organization of theaters and opera houses, and on the health of singers, musicians, and dancers.

The multidisciplinary approach in the writing of the book makes it accessible to different categories of audiences: it is of interest not only to experts in the field but also to all those who are fans of opera and wish to explore the connections between opera and medicine, and between opera and the brain.

Merate (LC), Italy
Monza (MB), Italy

Lorenzo Lorusso
Michele A. Riva
Vittorio A. Sironi

Contents

Part I

ACT I Medicine Goes at Opera

1 Pathology and Lyrics: Diseases and Illnesses in the Opera

Vittorio Alessandro Sironi and Andrea Scarduelli

1.1 The Sanitary Prism of an Era

Between the eighteenth and twentieth centuries, musical melodrama—as well as literature and art in general—was able to be a faithful reflection of society and time, a "window" to watch the evolution of knowledge in the medical field. In this context it is therefore inevitable that even the health aspects (the most frequent and evident diseases, the sometimes contradictory and often ineffective work of doctors and pharmacists, the limits and contradictions of a medicine unable to correctly recognize and effectively treat pathological conditions, the variegated world of charlatans and popular barkers) appeared in operas [1–3].

Above all, the pathological manifestations that struck, impressed, and mostly involved the people were the privileged themes that were included in the plot of the opera and that appeared on the theater scene [4–6].

Due to their spectacularism, especially neuropsychiatric diseases are those that are most commonly present in operas. Although there is no lack of dramatic pictures that portray spinal dimorphisms and deformities, recurrent infectious diseases or other pathological conditions determine a sensory handicap (such as blindness or deafness) or important sequelae (such as those resulting from injuries sustained in war or duel), situations that constitute the effective portrait of the health condition of the past centuries as perceived and experienced by the people [7–9].

V. A. Sironi (✉)
History of Medicine, School of Medicine and Surgery, University of Milano-Bicocca, Monza, Italy
e-mail: vittorio.sironi@unimib.it

A. Scarduelli
Teatro Belloni – Barlassina (Monza), Barlassina, Italy

L. Lorusso et al. (eds.), *Effects of Opera Music from Brain to Body*, Neurocultural Health and Wellbeing,
https://doi.org/10.1007/978-3-031-34769-6_1

1.2 Madness in Opera

The theme of madness is one of the most recurrent and the methods of scenic representation of this existential condition are a faithful mirror of the change in popular and scientific conceptions in this regard, thus allowing us to reconstruct cultural, social, and medical beliefs towards it [10–12].[1]

Madness has been present in the world of entertainment since the times of ancient Greece in the schemes of classical theater, then developing over time first in the medieval "festival of fools" and then in the "commedia dell'arte" [13]. "When melodrama was born and asserted itself in the seventeenth century—wrote Alberto Pellegrino—the so-called 'grand scene' takes shape on the stage, in which a singer or a singer has to deal with the performance of a musically very elaborate aria, in which it is necessary to demonstrate one's vocal resources and interpretative abilities [which] represented the moment of maximum expressive tension throughout the seventeenth and eighteenth centuries during which the character is [...] puts love passion first which often culminates with a striking manifestation of mad delirium for lost or lack of love" [14].

From *Il ritorno di Ulisse in patria* (1640) by Claudio Monteverdi to *Orlando* (1733) by George Friedric Handel, from *Iphigénie en Tauride* (1779) by Christoph Wullibald Gluck to *Nina, o sia La pazza per amore* (1789) by Giovanni Paisiello, madness often appears as a dream bewilderment or loss of identity as well as true mental illness. It is the exaggerated passions that lead to incorrect behavior and, often, individual, social, and moral fear becomes the expressive modality through which the lost psychic balance is manifested [15].

However, the greatest expressive drama occurs with the advent of Romanticism, a period in which literally and melodramatically we witness the triumph of passions and the exaltation of sentiment, with the Shakespearean rediscovery of the weight of events and destiny on human affairs. In the work, scenes of female madness prevail with women going mad after a love disappointment (as in the *Anna Bolena* in 1830 or *Lucia di Lammermoor* in 1835 by Gaetano Donizetti), while the less frequent male madness is more an expression of moral remorse derived from criminal acts (as in *Semiramide* in 1823 by Gioachino Rossini or *I masnadieri* in 1847 by Giuseppe Verdi) [16, 17].

Between the nineteenth and twentieth centuries in opera, compared to other pathological neuropsychic manifestations, the scenes of madness take on a particular relevance, both in relation to the evolution of medical thought in this regard that develops in the decades of transition between the two centuries and in particular with regard to the development of twentieth-century psychoanalysis. "The exaltation of passions, often uncontrollable and even obsessive, finds an outlet in madness—writes Alberto Pellegrino—which occupies a leading role in melodrama, so that in some opera scenes something morbid, neurotic, delusional circulates [...], A bundle of extreme and violent passions that converge in the 'scene of madness' which represents a culminating and even autonomous moment within the romantic

[1] A systematic list of operas in which madness is present in Chap. 5.

melodrama, becoming an opportunity for the performers to give free rein to pure virtuosity. In fact, madness, understood as a detachment from reality, justifies by the authors the assumption of a vocal language based on passages of agility and on ornaments, such as the 'warbling' or the 'floret', which give the measure of a strong totally exceptional emotion […]. Thus, a particular affinity is born between the human voice and some instruments such as the clarinet, the flute, the oboe because the musicians abound in this case the 'esplanade style' used in realistic passages, to adopt a 'flowery style' that represents better the detachment from reality determined by madness, giving free rein to the excited and impetuous singing" [14].

Spectacular (in the literal sense of the term) and deserving of careful analysis is the representation of madness in the three major Italian authors of nineteenth-century works: Gaetano Donizetti (1797–1848), Vincenzo Bellini (1801–1835), and Giuseppe Verdi (1813–1901).

Donizetti is the composer who addresses the theme of madness in a broader way. In the aforementioned *Anna Bolena* (1830), it represents the drama of this woman divided between the duty of queen next to Henry VIII and the call of her old love of her for Lord Percy. Imprisoned and sentenced to death Anna falls into a state of mad despair, upset by the loss of the crown and of life itself. In the opera *Il Furioso all'isola di San Domingo* (1833), referring to an episode of Cervantes' "Don Quixote," the maddened character is Cardenio, who has lost his reason for the betrayal of his wife and who finally succeeds to regain reason and the woman he loves [18] (Fig. 1.1).

In *Lucia di Lammermoor* (1835), a story centered on the hatred that divides two families, Lucia, who has sworn eternal love to Edgardo, is forced by her brother out of simple political opportunism to marry the noble Arturo. During the ceremony, Edgardo bursts onto the scene accusing Lucia of having betrayed their oath of love. Lucia, distraught, kills Lord Arturo and then returns to the party hall in the throes of a madness that will lead to her death [19]. "This is the most famous 'scene of madness' of opera production of all time—comments Alberto Pellegrino—, a testing ground for the skill of opera singers because it requires a great virtuosic technique due to the extreme difficulties of vocal weaving, since the melodic charge oscillates continuously between the recitative, the mid-aria and the actual aria. Completely out of her mind, Lucia remembers her secret encounters with Edgardo and imagines celebrating her wedding through a series of extraordinarily intense arias (*Il dolce suono mi colpì di sua voce, Sparsa è di rose, Ah! L'inno suona di nozze!, Ardon gli incensi*) which end with the death of the girl […] In no other melodrama do we find that an overwhelming amorous passion can so suddenly turn into an irreparable tragedy" [14]. The Donizetti representation combines the romantic rediscovery of Shakespearean-inspired feelings (William Shakespeare, 1565–1616), English playwright capable of characterizing and analyzing the feelings and emotions of people, humble or noble as they were) with the first scientific and medical evidence inspired by Pinel (Philippe Pinel, 1745–1826), the French doctor who first recognized mental illness as a medical problem, distinguishing it from other forms of social exclusion) of madness as an illness and not as an element of moral alienation. Regarding this scene, the critic Rodolfo Celletti points out: "This immaterial, yet intensely

Fig. 1.1 Front of page of *Il Furioso all'Isola di San Domingo* al "Teatro alla Scala", Milan in 1843—Spring

emotional way of describing the desperation of a being whom love has led to murder and madness, blossoms into hallucination when the echoes and calls that come in Lucia's upset mind they dissolve in the wordless song of very daring vocalized passages iterated by the glassy and feeble sound of a flute also engaged in spasmodic virtuosity" [14, 20].

Another work by Gaetano Donizetti, *Linda di Chamounix* (1842), deals with the theme of madness unleashed by betrayed love, a theme that had already been dealt with in *I puritani* (1835) by Vincenzo Bellini, another great composer who is not indifferent to the theme of madness. Indeed, he is the first of the great romantic musicians to tackle the theme of diversity, as will be said about the representation of oneiric thought found in his melodrama *La sonnambula* (1831) [21, 22].

Giuseppe Verdi gives a different, more dreamlike, and cathartic imprint to the theme of madness. His female characters, sopranos all outlined in strong colors, accentuate aggression and passion in the song, because madness becomes a nightmare, fear and obsession, and, at the same time, an outlet and catharsis for one's sins. In *Nabucco* (1842) the protagonist, King Nabucodonosor, returned to his homeland victorious after having subdued Israel, in the face of his boundless pride that leads him to become blasphemous by proclaiming himself God, is punished by the Divinity who makes him lose the throne, and makes him leave hindsight, upset by a delirium supported by the vision of terrible ghosts (Fig. 1.2).

Imprisoned in madness and tormented by shocking nightmares, he will hardly find the reason and strength to regain the throne by promising freedom to the Jews. In this case, madness is first divine punishment to strike the presumption of a king who wants to become God, but then also an instrument to rediscover royal reason and wisdom [23].

The theme of madness returns overwhelmingly in *Otello* (1887), where the Moor of Venice is devoured by a pathological form of jealousy that leads him to murder an innocent and destroy himself and where even Iago, incarnation of evil, appears to be affected by an equally pathological form of envy accompanied by an irrational hatred that leads him to destroy three existences. In the lyric drama, two other key characters experience existential disintegration: Desdemona and Otello. In fact, the identity of Desdemona, an intelligent woman and respectful consort, crumbles under the blows of Othello's madness. In the scene full of pathos of the dramatic night of his assassination, Desdemona "feels her existence heading towards a tragic destiny, announced also by the song of the Ave Maria, [...] in the sign of an inner

NABUCODONOSOR

DRAMMA LIRICO

IN QUATTRO PARTI

DI

Temistocle Solera

DA RAPPRESENTARSI

NELL' I. R. TEATRO ALLA SCALA

IL CARNEVALE DEL 1842.

Fig. 1.2 The first Front page of the "Nabucodonosor" for La Scala—Milan, 1842

fear, of a loss, of a sense of impending death that recall ancient and ancestral premonitions, which make the psychology of this female character particularly fragile, who accepts […] her impending destiny, victim of the reveries of a man who has transformed the love feeling into a wild and unstoppable nightmare [because] Othello's irrationality and morbid imagination prevail over love and, in the grip of blind jealousy, transforms his faithful wife into an adulterous sinner who must be absolutely annihilated" [23, 24].

Finally, another Verdi opera inspired by the great genius of Shakespeare represents another moment of dreamlike suggestion capable of involving, for its evocative manifestation, the curiosity and attention—almost morbid—of the general public, who love to get involved in the representation of love affairs with disturbing and mysterious implications. In *Macbeth* (1847) he depicts a woman (Lady Macbeth) obsessed with memories of her and her husband's crimes who manifests her nightmares and fears of her in sleepwalking [25].

1.3 Sleepwalking and Other Neurological Manifestations

Sleepwalking and sleep disorders are, after madness and at the same time sometimes madness, the other "mysterious" and "fascinating" neurological conditions most represented in the work. Considered for centuries the expression of a diabolical intervention, this "wandering in sleep," in which automatic motor acts (such as walking) are often accompanied by more complex gestures and the emission of sounds and words, only in the early twentieth century did we begin to understand how in reality this neurological manifestation was a complex sleep disorder (parasomnia) often not easy to diagnose [25]. In the romantic era these behavioral manifestations aroused interest and curiosity, thus making it understandable how it could be an element of great narrative effect in melodrama. The absolute peaks of these representations can be found, as already mentioned, in Giuseppe Verdi's *Macbeth* (1847) and Vincenzo Bellini's *La sonnambula* (1831) [22].

In the "sleepwalking scene" of Verdi's opera, the presence of a visionary lady suspended in an unreal fixity contrast with the realistic presence of the lady-in-waiting and the doctor, who observe with incredulous attention the unfolding of the drama of a woman's dream madness who sinks into a depression that reveals the psychotic origin of his thirst for power and his inherent inability to live. The diabolical woman who inspires horrendous crimes is transformed into a wide-eyed "human larva" who gives free rein to her madness, a logical outlet for her sense of remorse and guilt. Verdi's vision overlaps with the Shakespearean work that inspired it, in which sleepwalking is an expression of the incessant anxiety that assails those who feel guilty and sinful [14, 26, 27].

Amina's sleepwalking in Bellini's masterpiece is based on a more "modern" vision of the dream phenomenon understood as an innocent manifestation of a sleep disorder and not as a pathological expression of a "moral evil" and a guilt to be expiated. The young woman arouses immediate understanding and sympathy in the public and the revealing diagnosis, which dispels all doubts, is by Count Rodolfo.

Although he does not know the healing arts of medicine, he represents the symbol of a lucid and rational vision of a natural event, even if it is not frequent, not a disease, but an unusual and innocent mode of behavior, thus unreservedly chasing away the superstition popular beliefs that saw in this manifestation the expression of an evil influence [28].

Finally, albeit more rarely, we find in some works more or less explicit references to neurological disorders such as vascular disorders at the origin of motor deficits (hemiparesis and hemiplegia). Epileptic seizures are also not absent in the melodrama: in *Otello* (1887) by Giuseppe Verdi and in *Ivan IV*, composed in 1862 (first performance 1951) by Georges Bizet. There is also the presence of dementia in recently represented works inspired by the story of King Lear by William Shakespeare: *King Lear* composed between 1880 and 1885 (first performance 2009) by Antonio Cagnoni and *Lear* (1978) by Aribert Reiman and *Re Lear* (1850) by Giuseppe Verdi on libretto of Antonio Somma. We cannot forget *The Lion's Face* (2010) by Elena Langer and Glyn Maxwell [29].

1.4 Tuberculosis

Tuberculosis, an infectious disease widespread everywhere and known since ancient times, was another frequent "medical emblem" in art and in particular in theatrical and lyric works of the Romantic period [6]. "Consumption"—a common name with which the pulmonary form of tuberculosis was designated—is a disease with a wide operatic resonance. It constitutes a central element in the construction of the plot of the opera *La traviata* (1853) by Giuseppe Verdi. How can we forget the figure of Violetta Valéry, who literally dies on stage both for her love for Alfredo and for her illness which irreversibly compromises her health [7, 30]?

Consumption, the subtle ache, was the disease of the poor, of prostitutes (a counterpoint to the equally feared "romantic disease," syphilis). And it was the literary pathology that most exhausted the sinners, such as Violetta in *La traviata* or Mimì in *La bohème* [31]. Peculiar (narrative) retaliation: the red of sin on the one hand and the red of blood as punishment. On very white skin, like the bloodless one of many literary protagonists struck by tuberculosis [32] in a world that, if possible, is even more rotten than the disease: rancor, erotomaniacs, avenging relatives, and family poisons. The nineteenth-century realist novels and the melodramas inspired by them have often set consumption in a universe that is inherently vicious, as if this were the apotheosis of perdition, the gaze on the abyss of romantic painters, the last act of an extreme world and exhausted [33].

La traviata is a work full of very profound meanings. Great feelings are perceived in the melodrama, but also changes, sacrifices of love, and erroneous appearances, like the one that would reveal Violetta's frivolous personality, who instead in her dramatic song shows unexpected courage and clarity about the fate that awaits her. Despite her very young age, the protagonist is aware of her path as a sick person and of the death she is going to meet. She thus turns to God, accompanied by a melody that constantly follows her in the memory of worldly life, to the experience of the festivities of the high-ranking society, but also to the love of Alfredo [34].

The medical connotation is linked not only to the consumption caused by the consumption that affects the lungs, but also by the pallor of the skin that denotes the anemia that accompanies the pathological manifestation and expresses—as well as the weak voice—the progressive slow dying out of existence.

La bohème (1896) by Giacomo Puccini is also marked by the trend of this pathology. The time span of the story (one year) is governed by the development of Mimì's disease: from the initial manifestations of symptoms to the tragic epilogue of death, testifying to an inexorable disease that leaves no way out for the patient [35].

Puccini's opera was born from a challenge between Giacomo Puccini and Ruggero Leoncavallo, who compete to simultaneously write two works of the same name taken from the same source. After more than a century, however, Puccini's work is still among the most popular in the world, while that of Leoncavallo, with the same title and the same plot, has never met with much success [36].

1.5 Spinal Dysmorphisms and Deformities

Some famous romantic works contain references to spinal deformities. In *Rigoletto* (1851) by Giuseppe Verdi, the protagonist of the same name is "a hunchback," which is the transposition into Verdi's melodrama of the best-known literary character with a spinal deformity: the hunchback of Notre-Dame. The protagonist of Verdi's opera could be a subject suffering from severe adolescent idiopathic scoliosis, as some scholars have hypothesized [37] and the melodrama takes up and interprets the nineteenth-century attitude towards deformity: the radicalization and exclusion of hunchbacks from the context social and community, relegating them to marginalized figures. Also, in a previous opera, *Esmeralda* (1836) by Louise Bertin, taken from the same literary source, the figure of Quasimodo highlights the social discrimination against this deformed subject, perhaps suffering from Recklinghausen's neurofibromatosis (Fig. 1.3).

Their behavior is described as brutal and evil. Although both Rigoletto and Quasimodo show intense love, at the end of each work, they suffer a serious defeat with the loss of this great love. For nineteenth-century society, people with deformities and disabilities are considered worthless in the human and community sphere. And this is what the works faithfully reproduce in the elaboration of their plots [38].

1.6 Sensory Deficits

In *Iolanta* (1892), the last work of Pyotr Il'ič Čajkovskij, the drama of Iolanda of Anjou is represented, blind from birth (which she does not know), who, after a troubled and complex existential love, get your sight back. It is a singular work, in which a serious sensory deficit is represented—blindness in fact—which however

Fig. 1.3 Théâtre de l'Académie Royale de Musique where *La Esmeralda* premiered in 1836, in Paris

constitutes a handicap that is ultimately overcome more for the action of feelings than for the care of doctors [39].

Although it is perhaps the most striking example, it is not the only reference to blindness and visual impairment found in her works. A systematic review made it possible to highlight these situations (congenital blindness, linked to trauma or unknown causes) in 38 works written and represented in the last three centuries [40].

An equally unique and particular representation at the Municipal Theater of Piacenza concerning deafness was recently performed in the opera *Il sordo. Sogno d'amore in una notte di S. Silvestro* (2020), with the music of Paolo Marcarini and the lyrical interpretation by Leo Nucci, who staged the drama of the great composer Ludwig van Beethoven in hearing the complete deafness approaching that would have prevented him from hearing his compositions [41].

1.7 Wounds

Wounds (from sharp blades or guns) often recur in melodrama, indirectly or directly. In *La forza del destino* (1862) by Giuseppe Verdi, Don Carlo di Vargas, wounded by a gunshot, is fortunately saved by the surgeon who manages to extract the bullet and declare the operation out of danger. However, there is not always a happy ending. In Umberto Giordano's *Fedora* (1898), the surgeon called to help Count Vladimir Andrejevich, betrothed to Princess Fedora Romazov, seriously injured in mysterious circumstances, was unable to save his life. We also mention *Tristan und Isolde*

(1865) by Richard Wagner and, recently, *Fallujah* (2016) by composer Tobin Stokes and librettist Heather Raffo concerning the US-Iraq conflict in 2004 [42].[2]

1.8 Conclusion

The librettos of many of the most famous operas contain references to the disease which can thus constitute valuable sources of information on how different pathological conditions have been experienced and understood over time.

References

1. Carmody J. Doctors and opera. Med J Aust. 1991;155:783–4. https://doi.org/10.5694/j.1326-5377.1991.tb94038.x.
2. Willich SN. Physicians in opera--reflection of medical history and public perception. BMJ. 2006;333:1333–5. https://doi.org/10.1136/bmj.39056.464074.55.
3. Norio R. Medical profession in opera. In: Klockars M, Peltomaa M, editors. Music meets medicine. Helsinki: The Signe and Ane Gyllenberg Foundation; 2007.
4. Soriano JB. On doctors and their operas: a critical (and lyrical) analysis of medicine in opera. Chest. 2018;154:409–15. https://doi.org/10.1016/j.chest.2018.03.015.
5. Mason RB. Seeing medicine through opera glasses. CMAJ. 1996;154:921–3.
6. Hutcheon L, Hutcheon M. Opera, desire, disease, death. Lincoln: University of Nebraska Press; 1999.
7. Hutcheon L, Hutcheon M. Opera: the art of dying. Cambridge: Harvard University Press; 2004.
8. Vannoni G. A un dottor della mia sorte. Bologna: Pendragon; 2017.
9. Lorusso L, Franchini AF, Porro A. Opera and neuroscience. Prog Brain Res. 2015;216:389–409. https://doi.org/10.1016/bs.pbr.2014.11.016.
10. Verdeau-Paillès J, Laxenaire M, Stoecklin H. La Folie à l'Opera. Paris: Buchet/Chastel; 2005.
11. Sala E, Ashbrook A. Women crazed by love. An aspect of the romantic opera. Opera Q. 2003;10:19–41.
12. Lorusso L, Porro A. Opera and neuroscience: a historical approach and its relevance today. In: Colombo B, editor. Brain and art. From aesthetics to therapeutics. Cham: Springer; 2020. p. 81–94.
13. Pugliese RM. Il topos della follia nell'opera. Prospettive critiche. In: Sala E, editor. Mad Scene and Songs. Quaderno delle notti Malatestiane. Rimini: Raffaelli; 2002. p. 115–29.
14. Pellegrino A. Il tema della follia nell'opera lirica. Lettere dalla Facoltà. Bollettino della Facoltà di Medicina e Chirurgia dell'Università Politecnica delle Marche. 2012;15(4):29–35.
15. Castelvecchi S. From "Nina" to "Nina": Psychodrama, Absorption and Sentiment in the 1780s. Camb Opera J 1996; 8: 91–112.
16. Gerber MA. Power in madness: a critical investigation into the musical representation of female madness in the mad scenes of Donizetti's 'Lucia' from Lucia di Lammermoor (1835) and Thomas's 'Ophélie' from Hamlet (1868). Thesis. Pretoria: University of Pretoria, South Africa; 2015.
17. Sharma TG. Lucia's Liberation: Feminine Madness in Donizetti's Lucia di Lammermoor. Thesis. Wellesley, Wellesley College; 2020.
18. Peschel E, Peschel R. Donizetti and the music of mental derangement: Anna Boleyn, Lucia di Lammermoor, and the composer's neurobiological illness. Yale J Biol Med. 1992;65:189–200.

[2] See also Chap. 8.

19. Harris RN. Going mad: a comprehensive study of the mad scene in Gaetano Donizetti's opera, Lucia di Lammermoor. Honors Theses. Arkadelphia: Ouachita Baptist University; 2013.
20. Celletti R. Storia dell'opera italiana. Milano: Garzanti; 2000.
21. Pugliese RM. The origins of "Lucia di Lammermoor's" cadenza. Camb Opera J. 2004;16:23–42.
22. Riva MA, Sironi VA, Tremolizzo L, Lombardi C, De Vito G, Ferrarese C, Cesana G. Sleepwalking in Italian operas: a window on popular and scientific knowledge on sleep disorders in the 19th century. Eur Neurol. 2010;63:116–21.
23. Cambioli L, Bava M, Bellelli G, Clerici M, Cesana G, Riva MA. "Nabucco" by Giuseppe Verdi: a case of delirium in an Italian romantic opera. Eur Neurol. 2017;77:180–5.
24. Rusbridger R. Projective identification in Othello and Verdi's Otello. Int J Psychoanal. 2013;94:33–47.
25. da Mota GM, Nardi AE. Lady Macbeth's night walking with dissociative symptoms diagnosed by the first sleep medicine record. Front Psychol. 2021;11:563773. https://doi.org/10.3389/fpsyg.
26. Hibberd S. Dormez donc, mes chers amours': Hérold's "La Somnambule" (1827) and dream phenomena on the Parisian lyric stage. Camb Opera J. 2004;16:107–32.
27. Finger S, Sironi AV, Riva MA. Somnambulism in Verdi's Macbeth and Bellini's La Sonnambula: opera, sleepwalking and medicine. Prog Brain Res. 2015;217:187–215.
28. Scuderi B. Vincenzo Bellini's La sonnambula: a clinical case on stage. J Hist Med. 2019;31(2):375–90.
29. Fuller D. Dementia in the opera: the Lion's face. Opera Q. 2011;27:509–21.
30. Barnes AC. La Traviata. JAMA. 1969;208:93–6.
31. Murray JF. Willing suspension of disbelief: new roles for Violetta and Mimi. Am J Respir Crit Care Med. 2013;187:465–7.
32. Abigail AL, Hoenig LJ. Enjoying opera. Dermatological style. JAMA Dermatol. 2016;152:327.
33. Hepokoski J. Genre and content in med-century Verdi: "Addio, del passato" (La Traviata), Act III. Camb Opera J. 1989;1:249–76.
34. Edgecombe RS. Some observations on the love: duets in *La traviata*. Opera Q. 2005;21:217–21.
35. Arosio L, Pellegrini L, Cesana G, Riva MA. The death of Mimì in Puccini's La Bohème: not an ordinary tubercular heroine. Lancet Respir Med. 2017;5:617–8.
36. Bianconi LG. Introduzione a Musica e Medicina. Bull Sci Med. 2010;182:3–8.
37. Andreotti M, Caruso G, Massari L, Riva MA. Spinal deformities in romantic operas. Spine. 2018;43:1617–8.
38. Fardon D. Spine in the arts. The hunchback in the opera. Spine J. 2002;2:158–9.
39. Kors S. Iolanta. Peter Ilyich Tchaikovsky. Opera Q. 1997;13:198–201.
40. Aydin P, Ritch R, O'Dwyer J. Blindness and visual impairment in opera. Eur J Ophthal. 2018;28:6–12.
41. See online: Successo dell'opera "Il sordo" al Municipale in www.ilnuovogiornale.it/archivio-articoli/notizie-varie/4920-successo-dell-opera-il-sordo-al-municipale.html visited on 18 May 2022.
42. Worth EJ. The changing role of the physician in opera. Opera Q. 1993;10:143–56.

2 Physicians, Charlatans, Healers, and Apothecaries in Operas

Vittorio Alessandro Sironi

2.1 The Physician Figures

Although often outlined in a fixed stereotype, the figure of the doctor appears with a certain frequency in works between the eighteenth and twentieth centuries: often the protagonist or in more secluded roles of supporting actor at other times. In the context of the historical evolution of the opera, these figures emerge as characters with a precise and sometimes complex physiognomy, representing characters familiar to the general public, capable of always arousing lively, often conflicting reactions: high admiration or great contempt. On the other hand, being all figures involved, in various capacities and in different ways, towards health (physical, mental, and often even sentimental) makes them fundamental in any case in the vicissitudes related to the narrative plot, even independently of the artistic role, that they play in the context of the opera [1].

Some aspects concerning the image that emerges of these figures in the opera literature allow us to analyze their essential features. A first important consideration is that relating to the professional skills of these "health workers," often described through a satirical and comic characterization, moreover well rooted in the theatrical tradition. The inexperience of doctors, often completely unable to assess the actual conditions of a sick person, is well present in the operatic literature, which offers numerous examples of doctors who diagnose and issue prognoses immediately and comically denied in the course of events [2].

In Gioachino Rossini's *Il Viaggio a Reims* (1825), "the doctor Don Prudenzio comes into action when the garrulous and frivolous countess of Folleville faints

V. A. Sironi (✉)
History of Medicine, School of Medicine and Surgery, University of Milano-Bicocca, Monza, Italy
e-mail: vittorio.sironi@unimib.it

L. Lorusso et al. (eds.), *Effects of Opera Music from Brain to Body*, Neurocultural Health and Wellbeing,
https://doi.org/10.1007/978-3-031-34769-6_2

from pain [and], rushing to claim only the right to give her treatment, judging the woman in life threatening, convinced of the extreme gravity of the case, first ruling [that] 'vital functions are suspended', then, as soon as he pronounces the ominous prognosis 'he will die', the sick woman regains consciousness, [for which] she does not find then better [...] than issuing the belated diagnosis 'was a syncope', only managing to relieve the sarcasm of bystanders" [3]. A similar situation is found in Ferruccio Busoni's *Arlecchino* (1917). Leandro, wounded by Harlequin in a duel, lies on the ground examining him and, without any hesitation, Doctor Bombasto declares him dead, but then, when the wounded man recovers and gets up, he blames the unexpected resurrection to the supernatural intervention of the devil. A similar situation is found in Paisiello's *Il Barbiere di Siviglia*, a work in which Doctor Bartolo also clamorously misdiagnoses Don Basilio, declared seriously ill while in reality he is in good health [4].

Similar incompetence is shown by the doctor Don Ipocrate (*nomen omen*!) in *Il medico parigino* or *L'ammalato per amore* (1792) by Gennaro Astarita, who predicts a few days of life for Sofonisba's suitor Knight (the woman the old doctor actually fell in love with) that in reality he pretended to be ill just so as not to make him suspicious. Maestro Spinelloccio in Giacomo Puccini's *Gianni Schicchi* (1918) also belongs to this category of "incapable doctors" of which operas abound (Fig. 2.1).

Fig. 2.1 Poster, by Italian illustrator Leopoldo Metlicovitz (1868–1944), advertising Puccini's opera *Gianni Schicchi*, published by G. Ricordi, Milan, in 1918–1919, in coincidence with the Italian premiere on January 11, 1919

He "enters the scene at a very precise moment in the work—writes Alberto Tanturri -: when Schicchi conceived, but not yet revealed to the relatives of the deceased Buoso Donati, his plan to replace the dead man, with the aim of dictating, taking advantage of his qualities as an imitator, a new testament that takes into greater account the expectations of relatives. The sudden arrival of Spinelloccio, still unaware of Buoso's death, risks disrupting Schicchi's Machiavellian plan; but the latter, having entered the dead man's bed, counterfeits his voice [and] Spinelloccio is satisfied with evaluating the condition of the patient only by his voice [...] and, reassured by his relatives, is pleased with the efficacy of the therapies he has prescribed, [...] Attributing the healing to the merits of the Bolognese medical school" [3]. The doctor intriguing intrigues had previously already been sketched in *Don Pasquale* (1843) by Gaetano Donizetti, a work in which Dr. Malatesta devises a cruel joke which, through an intricate stratagem, induces the elderly patient and friend Don Pasquale to give up forever for marriage purposes, also to safeguard their mental health [5].

The ineptitude of doctors and the poor consideration of their professionalism in the operatic context is even more evident in the consultation, that is, when there is a group of doctors gathered at the bedside of the patient who dispute among themselves in relation to diagnosis and therapy. Then processes of competition are triggered that are completely unrelated to the interests of the patient, since the interest of doctors is exclusively that of gaining consideration by showing off their doctrine through the use of Latin, an expressive modality that characterizes in a particular way the figure of the doctor in melodrama with a double purpose: to allow the doctor to emphasize his own cultural superiority, but also to mask the emptiness and inconsistency of his own doctrine unable to realize with bombastic expressions and words, effective interventions [2].

Emblematic, in the work of Ermanno Wolf-Ferrari *L'amore medico* (1913), the consultation to which the protagonist, Lucinda, a girl whose only evil is the deep melancholy due to the paternal refusal to marry her beloved Clitandro, is subjected. The doctors called to the woman's bedside expressed different and conflicting opinions. The first, Tomes, declares that the sick woman suffers from "*sanguinis ebollitio*," and therefore, she needs a bloodletting. The second, Desfonandres, diagnoses a "*gastrici congestio*" to be treated with emetics. The third, Macroton, finds a "*vaporum fumigatio*" and prescribes a purgative. The fourth, Bahis, considers the woman congested with "*humorum cerebri glutinorum*" to be cured with adequate herbal teas. Disagreeing on diagnosis and treatment, the four doctors mutually insult each other without proposing a valid curative solution for the patient. A similar situation is found in the work of Pasquale Sogner *Privazione genera desiderio, ossia moglie libera e collo torto* (1816) in which, around the bed of the fake sick Dorando, the consultation takes place between Pasquale, a doctor, and his brother Bernardo, a surgeon, who, due to the interpretative dissent on the reasons for the patient's malaise, exchange insults and almost come to blows. "It was actually a curious scene," comments the house servant Andrea talking to the other servant Marina who, in turn, notes that in this way "the sick person has not yet known what is wrong with him." To which Andrea wisely concludes: "This usually happens when there are more doctors. They contradict each other, and when they have what is badly fixed, they find the poor sick person already dead" [3, 6].

Lapillissian and tragic truth is that of the inconclusive litigiousness of doctors which appears in the eyes of the simple and is not perceived by skilled doctors, therefore the subject of frequent criticism in this regard, as in Pietro Generali's *Il medico e il ciabattino* (1918) or in *Cendrillon* (1899) by Jules Massenet or in *L'amour des trois oranges* (1921) by Sergej Prokofiev or lastly, as the title explicitly states, in *L'ammalata e il consulto* (1837) by Giuseppe Malusardi, in which a recurring plot is proposed. Elide, in love with Gerardo, clashes with her uncle who would instead want to marry her to the mature Don Protopio, Gerardo's father. The girl then pretends to be mad and at her bedside, summoned by the village farmers, 30 doctors arrive who, on the basis of the symptoms observed (irregular pulse, dry mouth, lack of appetite, and pallor), diagnose an unfortunate prognosis of imminent death. Except then, when the patient reveals her deception by getting out of bed in excellent health, they change their judgment by declaring that the girl "has lost synderesis," thus using a sophisticated and cryptic word to recover their compromised and fallacious clinical reliability (Fig. 2.2).

2.2 The False Doctors, the Healers, and Charlatans

"Greed, ignorance, arrogance, nonsense, litigation constitute the poles within which the prevailing image of the followers of Aesculapius in the world of melodrama is contained. The doctor's professional arrogance, his refined clothing and his studiedly hermetic speech hide the emptiness of a baggage of notions that is useless. The doctrine of the doctor is a useless arsenal of dogmas and precepts devoid of any practical efficacy." This merciless portrait of the figure of the doctor in the opera drawn by the critic Alberto Tanturri [3, 7, 8] explains the fact that often in the works health "knowledge" is entrusted to "false doctors," perhaps more reliable and suggestive figures than those of "true doctors," or to "hybrid individuals," well characterized and such as to sometimes represent the main character of the opera itself. Such is Figaro, protagonist of *Il barbiere di Siviglia* (1816) by Gioachino Rossini, able to perform functions as a doctor much more than the homonymous character in the playful drama by Giovanni Paisiello with the same title (1782) or that of *Le nozze di Figaro* (1786) by Wolfgang Amadeus Mozart. In fact, speaking of Don Bartolo, he defines himself emphatically "his barber, his surgeon and his apothecary," quick and skillful "to purge, bleed" [7, 8], just as in the Rossini opera he qualifies even more precisely as a "barber, surgeon, botanist, apothecary, veterinarian, housekeeper," making it clear to Don Bartolo—who defines himself as a doctor, emphasizing in a famous aria as "a doctor of my lot"—that bleeding and purging were tasks more than a barber than a doctor [3, 7], to the point that to treat the house servants, you do not need a doctor but a barber to dispense "opio and sneeze" [3].

False doctors are portrayed as cheating charlatans no less skilled than licensed doctors, masks created for the purpose of refined disparaging intent upon a powerless and vain medicine, as are its official representatives. In the melodrama, itinerant

Fig. 2.2 Cendrillon. Poster, by Émile Bertrand (1842–1912), for Jules Massenet's *Cendrillon*, advertising the première performance at the Théâtre National de l'Opéra-Comique, Paris (1899)

practitioners charlatans were also tooth pickers and bone tanners, real "health merchants," who were recognized with the great merit of offering miraculous cheap remedies in the squares, making accessible to all what was the monopoly of a few: the possibility to heal [2].

In the opera *Crispino e la comare* or *Il medico è la morte* (1850) by Luigi and Federico Ricci, the humble Venetian cobbler Crispino Tacchetta, to relieve his disastrous financial and human situation, accepts the advice of a mysterious and unidentified "comare" (a woman dressed of black that appears in his presence that is none other than death) that invites him to pass himself off as a doctor, also suggesting an infallible way to issue exact prognoses: his presence (visible only to the eyes of the "fake doctor") next to the patient it indicated the next death, while his absence a positive resolution of the pathology. It is the desire to punish "certain medical doctors" for their "superb asinity" that pushes the woman to help the illiterate and ignorant cobbler, thus becoming, in the Ricci's melodrama, the emblem of

CRISPINO E LA COMARE

LIBRETTO FANTASTICO-GIOCOSO

DI FRANCESCO M. PIAVE

MUSICA

dei Fratelli

LUIGI E FEDERICO RICCI

ESPRESSAMENTE COMPOSTA PEL TEATRO

GALLO A SAN BENEDETTO

NELLA STAGIONE

DI CARNOVALE E QUADRAGESIMA

DEL 1849-50

VENEZIA

NELLA TIPOGRAFIA GASPARI.

Fig. 2.3 Title page of the libretto for "Crispino e la comare," Venice 1850. By brothers Luigi Ricci (1805–1859) and Federico Ricci (1809–1877), librettist Francesco Maria Piave (1810–1876)

distrust towards "real doctors", Wise only in empty words devoid of a logical and curative sense (Fig. 2.3).

A similar figure had already been sketched in the first aforementioned work by Pietro Generali *Il medico ciabattino* (1819), in which the protagonist, the cobbler Prospero, despite not qualifying as a doctor, faced with the inconclusiveness and ignorance of the doctors gathered at the bedside of Donna Claudia, proudly declares how, even in the absence of specific instruction, he is able to heal the sick using a little simple sneezing powder, something that the wise doctors cannot (Fig. 2.4).

False doctors, facilitated in their effective sanitary action by the fact that they are also faced with false patients, are found in Georges Bizet's opera *Le docteur miracle* (1857) and in that of Charles Gounod *Le médicin malgré lui* (1858), melodramas in which the plot, as in other librettos of the time, is recurring. In this last work the rough woodcutter Sganarelle, pretending to be a doctor, prescribes a therapy based on bread crumbs dipped in wine, which is the diet, to Lucinde, pretending to be

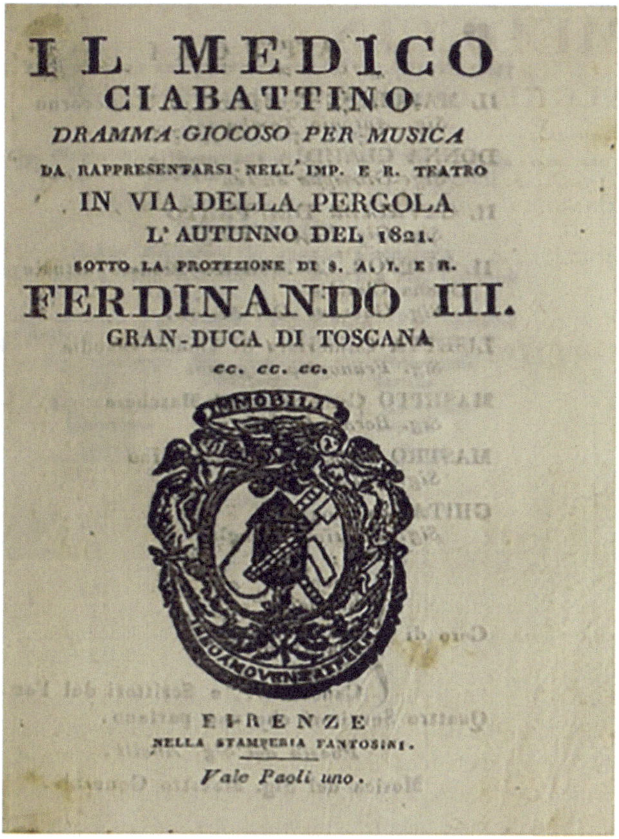

IL MEDICO
CIABATTINO
DRAMMA GIOCOSO PER MUSICA
DA RAPPRESENTARSI NELL' IMP. E R. TEATRO
IN VIA DELLA PERGOLA
L' AUTUNNO DEL 1821.
SOTTO LA PROTEZIONE DI S. A. I. E R.
FERDINANDO III.
GRAN-DUCA DI TOSCANA
cc. cc. cc.

FIRENZE
NELLA STAMPERIA FANTOSINI.
Vale Paoli uno.

Fig. 2.4 Front page of the "Il Medico ciabattino" (1821) by composer Pietro Generali (1773–1832)

silent because his father Géronte opposes his marriage to the young and destitute Léandre, typical of parrots, talkative animals par excellence (Fig. 2.5).

The treatment has immediate effective effects, not so much due to the virtues of the therapy, but rather because the opposition to marriage is less when Leandre finds himself receiving a rich inheritance. A similar story also takes place in the first mentioned work, where the love of Laurette and Silvio is hindered by her father, the Podestà of Padua. Silvio, pretending to be a doctor, succeeds with a subterfuge to make the Podestà believe that he has been poisoned and that he will be able to avoid his death if he will grant him the hand of his daughter ("si mihi filiam accordat in matrimonium").

An interesting "variation on the theme" concerning the mask of the fake doctor can be found in Mozart's *Cosi fan tutte* (1790), where the medical disguise of the servant Despina is functional to the plot hatched by Don Alfonso to demonstrate to his two friends Ferrando and Guglielmo that their girlfriends, Dorabella and Fiordiligi, respectively, are ready to betray them. Staged a false departure of the two

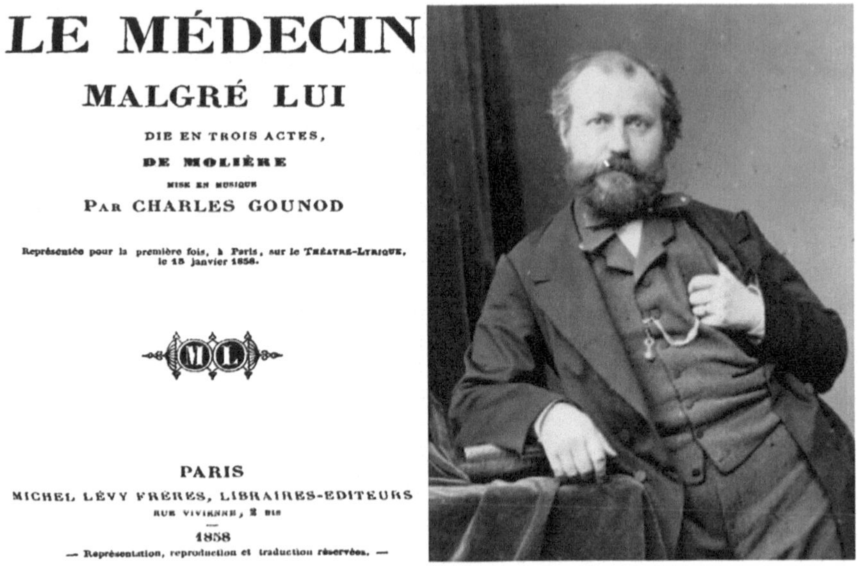

Fig. 2.5 Front page of the "Le Médicin malgré lui" (1858) by Charles Gounod (1818–1893) (portrait of 1860) from Bibliothèque National de France, Département de Musique, Paris

young men, Don Alfonso makes them disguise as Albanian officers and introduces them into the house of the two women, who, even if at first in the face of the love offers of the two young men, react indignantly, when the two men stage a fake poisoning from which they are "saved" by the intervention of the fake doctor (the servant Despina). She pulls out a magnet with which she touches the head and the body of the two patients explaining "This is that piece of magnet, Mesmeric stone which had its origin in Alemagna, which was then famous there in France" and urging the two girls to support the heads of young people to encourage their recovery. "Describing this bizarre cure—writes Tanturri—Mozart is merely staging the therapeutic methods (at the time in great vogue among the wealthy classes) of the Swabian doctor Franz Anton Mesmer, although this time with intent not caustically disparaging, but rather, gracefully parody" [3, 9–11].

In fact, especially between the mid-nineteenth and early twentieth centuries, the stereotype of the doctor tends to change: no longer just insipid or fake, but also competent and compassionate, like Dr. Grenvil in Giuseppe Verdi's *La Traviata* (1853), who at Violetta's bedside, while aware of the serious condition of the patient and aware of the limits of medicine with regard to tuberculosis (the disease affecting the woman), he comforts her and inquires about the quality of his night's rest, willing to do everything possible to assist the sick, hiding from her even paternalistically (as was the practice of the time) the imminence of her death.

After Grenvil there will be other compassionate doctors who will accompany some sick people to the last moments of their existence. Like the doctor who witnesses the death of the protagonist in *Pelléas et Mélisande* (1902) by Claude

Debussy or doctor Javelinot who administers a sedative to the elderly prioress who is dying in the opera *Dialogue des carmélites* (1957) by Francis Poulenc.

However, this new version of the "competent doctor" remains an isolated figure, because that of the charlatan is still the prevailing mask. "Ambiguous figures of thaumaturges who are not sedentary but wanderers, [who] offered a type of goods and services halfway between the doctor and the apothecary, an expression] of a magmatic healing world, [because]" on the one hand they visited the sick, formulated diagnoses and they prescribed treatments, but on the other hand they also sold drugs [1–3].

In fact, the charlatan gave his best by producing long tales—often with a singular comic effect—to proclaim the therapeutic virtues of the medicines he sold as panacea. Often a single preparation which was attributed "miraculous" healing properties relating to a specified number of heterogeneous diseases. In the opera *Il medico burlato* (1790) by Pietro Guglielmi, the dancer Flacco, dressing the loaves of the charlatan Primasso, appears at the bed of the sick person claiming to be able to heal "those who suffer from sight, those with a severed arm, those who do not have lung 'Because' in Cairo, in less than three minutes, I made the asses talk and the dumb walkers."

> A key element in the success of the charlatan—Alberto Tanturri explains—must in any case be attributed to his strategy of placing himself on the same cultural level as his patients/clients. Unlike the doctor, he does not resort to the hermetic Latin phraseology that confuses the simple and leads them to distrust [and moreover] if the doctor hypocritically disguises his venality, the charlatan does not hide the 'commercial' nature of his services, but rather, with the techniques of a consummate barker, he emphasizes its convenience, [making] his listeners believe that for a miserable price he is offering them an unmissable opportunity [3].

Emblematic in this sense is the figure of Dulcamara in *L'elisir d'amore* (1832) by Gaetano Donizetti, who does not hesitate to magnify the product he wants to sell, affirming its amazing properties: "This is the marvelous odontalgic liqueur of mice and powerful destroyer bugs, whose authentic certificates, stamped, touch, see and read to each one I will do. For this specific sympathetic, prolific man of mine, a septuagenarian and valetudinary man, grandfather of ten children, still became [...] He moves the paralytics, sends the apoplectic, the asthmatics, the asphyxiated, the hysterics, the diabetics; he heals tympanitides, and scrofula and rickets, and even the liver pain that became fashionable" [1, 2].

The reference to "certificates" alludes to the fact that, while charlatans were able to win the trust of humble people with respect to doctors, they nevertheless needed from the latter specific authorizations and specific certificates relating to the effectiveness of their products, obtained of course with high remuneration to those who issued them. In addition to archival documents, direct evidence of this can also be found in the works themselves, as is evident in the melodrama *Quattro prigionieri ed un ciarlatano* (1832) by Pasquale Sogner and in *Il mercato di Malmantile* (1758) by Domenico Fischietti. In the latter work, the charlatan Rubicone, only by presenting to the governor of the small town the certificates that qualify him as a "great doctor" does he obtain permission to sell his drugs without paying any tax. Thus, the existence of a nonconflictual but complementary relationship and implicit complicity between doctors and charlatans appears explicitly. For the former, the granting (for a fee) of privileges to charlatans constituted a not insignificant source of income, while for the latter the certificates were the best way to legitimize their business compared to street beggars.

2.3 Bad Doctors

The opera house has not remained indifferent even to the figures of doctors with negative connotations. The most conspicuous negative effect is that offered by the Docteur Miracle in *Les contes d'Hoffmann* (1881) by Jacques Offenbach, which causes the death of the fragile Antonia—as she had previously done with her mother—forcing them both to sing even if suffering from lethal tuberculosis and a fatal hereditary cardiac arrhythmia. "Mask of a wicked and perfidious doctor—comments Tanturri—Miracle embodies the contemporary nightmare of medicine perverted to evil and can be considered an unconscious literary anticipation of the role played by "cursed' doctors in Nazi concentration camps" [3, 12].

Also in Alban Berg's *Wozzeck* (1923), the sinister figure of the doctor embodies the physician who uses the new discoveries of science not for the benefit of patients, but only for his personal glory. He performs aberrant dietological experiments on the protagonist, the poor soldier Franz Wozzeck who lends himself to acting as a human guinea pig only for economic reasons. The doctor has no regard for his health condition, fantasizing about how his experiments will give him stardom and scientific immortality. "Such a person—writes Tanturri—points out against the light the need for the new and more solid doctrinal acquisitions of medicine to be supported by equally solid ethical and deontological requirements" [3].

2.4 Surgeons and Apothecaries

It is rare in the operas to compare the figure of the surgeon, but always described—unlike that of the doctor—as that of a valid and trained health worker. In *La forza del destino* (1862) by Giuseppe Verdi, the surgeon plays a fundamental role in the work. By saving the life of Don Alvaro, wounded in battle, he allows Don Carlo di Vargas to recognize him as the kidnapper of his sister Leonora and the alleged murderer of his father. Even if his presence on the scene is very short, he in front of the injured person is outlined as a skilled professional. "The ball in his chest scares me," he says with diagnostic precision and prognostic concern. But then he manages to extract the bullet and declare the operation out of danger. Less success has in the *Fedora* (1898) by Umberto Giordano the surgeon Lorek, called at the beginning of the work to help Count Vladimir Andrejevic, betrothed to Princess Fedora Romazov, seriously injured in circumstances. His treatment was ineffective and the wounded man died in a few minutes, as he had unfortunately immediately predicted upon seeing the count, having immediately judged the wound. However, his precise competence and the readiness immediately declared to collaborate with Dr. Muller (doctor with whom he assists the wounded) appear to be admirable in trying to save the patient's life.

The figure of the apothecary is more varied and amusing overall, an object of satire and comedy linked to an often caricatured vision of this character, of which it is useful to underline the double connotation present in the melodrama, distant both temporally and professionally [13, 14].

The first work to remember is *Lo speziale* (1755) by Vincenzo Pallavicini, in which the protagonist Sempronio appears, from the very first bars of the work, disinterested in his profession, more inclined to reading newspapers in search of curious news to comment with friends and acquaintances present in the shop. For this he systematically entrusts the preparation of medicines to his apprentice Mengone, semi-illiterate and completely unsuitable for this delicate task, because he does not know how to interpret the writing of doctors, confuses the units of measurement, and often decides to administer drugs at his will. What emerges is a figure of an unprofessional apothecary who uses the shop more as a place for socializing than as a center for the production and sale of drugs [15]. Same subject and same title (because both are taken from a libretto by Carlo Goldoni) has the comic opera in three acts by Joseph Haydn, performed for the first time in 1768, in which, however, the figure of the elderly apothecary Sempronio, who aspires to marrying his pupil Grilletta, appears more focused on the amorous rivalry towards the young rival boys Volpino and Mengone. It will then be the latter who will stuff the beautiful Grilletta, leaving the apothecary and the other boy in trouble, at the mercy of the Turks who plunder the pharmacy [16–18].

Figures of apothecaries who are not very attentive to their trade and often used as effective comic pretexts for the entertainment of the public are found in other operas, especially German: *Les eaux de Merlin* (1715) by Jean-Claude Gillier, *The Alchemist* (1710) by George Friedric Handel, *Die Apotheke* (1771) by Christian Gottlob Neefe, *Der Alchymist* (1778) by Johan André, *Der Alchymist* (1779) by Joseph Schuster, and *Doktor und Apotheker* (1786) by Ditters von Dittersdorf and then adapted into English in 1789 in *The Doctor and the Apotechary* by Stephen Storace [16, 17].

A few decades later, the radical change in the figure of the pharmacist and his role in the shop at the beginning of the nineteenth century is clearly highlighted in the opera *Il campanello dello speziale* (1836) by Gaetano Donizetti. At the opening of the work, the apothecary Don Annibale Pistacchio marries the much younger Serafina and, once the celebrations are over, the servant Spyridon offers, in case of nocturnal calls, to prepare and sell medicines instead of the master. But the apothecary declines the offer, reminding the unwary servant to avoid serious mistakes for himself and unforgivable irregularities for both. "Would you like to do—she reminds him—like last month you gave that poor woman the corrosive sublimate instead of cream of tartar? That petty one, because of your ignorance, ran the risk of going to the other world. Your blunder has meant that the government has ordered all the apothecaries to sell their drugs in person in night time, under penalty of fine and arrest."

This resolute decision allows the young Enrico, in love with Serafina, to continually disturb the apothecary at night by taking on various transfers, with the (finally achieved) aim of preventing the consummation of the wedding. "Beyond the conclusion of the opera—comments Alberto Tanturri—, which fits into the traditional comic cliché of the old babion mocked by a younger rival in love, the libretto documents an important passage in the evolution of the profession of apothecaries [profiling] a professional framework that required him to be qualified and expert knowledge of the pharmacological properties of the substances he worked" [3, 17, 18].

In spite of this, there is no lack of other "satirical masks" made to wear by the pharmacist in works of the following decades. As in the Mirabolano apothecary, set in Venice in the seventeenth century, in which "la teriaca is done" in the aforementioned *Crispino e la comare* (1850) by the brothers Federico and Luigi Ricci, or in the more recent funny act *La smorfia* (1959) by Bruno Bettinelli, whose action takes place in the early twentieth century, in which the owner of an old pharmacy transformed into a "Chemical Laboratory" is a pharmacist who blindly believes in the cabal of dreams ("La Smorfia" Neapolitan), which is why he provides customers the numbers to play the lottery more than drugs for their ills, also because his belief – which he does not hesitate to declaim – is that "in three classes he distinguishes the pharmacopoeia in his drugs: the useless, the harmful and the lethal" [18] (Table 2.1).

2.5 Conclusion

Although stiffened in a fixed stereotype, the figure of the doctor appears quite frequently in opera between the eighteenth and early twentieth centuries. He usually belongs to a medium–high social status with a high standard of living, often highlighted by his clothing which, in addition to being an emblematic sign of the profession, also shows the unequivocal economic wealth he enjoys. The latter expression is more of the skillful self-protection capacity of one's own state than of professional capacity. Indeed, there is little recognition of the social value of his work and of his professional skills. Doctors, apothecaries, and charlatans, often united in a single not so noble vision of the art of treatment, will remain for decades among the most well-known and "exploited" characters in the lyric field to show the limits of a medicine still unable to act as a scientific discipline capable of addressing and solving health challenges.

The doctor is almost always unable to formulate a valid diagnosis and to propose effective therapies. Indeed, in melodrama medicine is often perceived and conceived more as the work of barkers than of skilled professionals. For this reason, the figures of charlatans who propose empirical remedies abound, often also as an alternative to the preparations of the apothecaries and the "courtly" therapies of the doctors. The doctrinal background of doctors is often less effective and efficient than the common sense that simple and humble people are able to put in place to resolve health situations. The consultation is emblematic of the incapacity and inconclusiveness of doctors (and the maximum target of satire), a situation in which several

Table 2.1 Health care figures in opera

Composer	Opera	Health care figures
Jean-Claude Gillier	*Les eaux de Merlin (1715)*	*Apotechary*
George Friedric Handel	*The Alchemist (1732)*	*Apotechary*
Vincenzo Pallavicini	*Lo speziale (1755)*	*Apotechary*
Christian Gottlob Neefe	*Die Apoteke (1771)*	*Apotechary*
Johan André	*Der Alchimist (1778)*	*Apotechary*
Joseph Schuster	*Der Achimist (1779)*	*Apotechary*
Ditters von Dittersdorf	*Der Doktor und der Apotheker (1786)*	*Apotechary*
Stephen Storace	*The doctor and the apotechary (1788)*	*Apotechary*
Federico and Luigi Ricci	*Crispino e la comare (1850)*	*Apotechary*
Gaetano Donizetti	*Il campanello dello speziale (1836)*	*Pharmacist*
Bruno Bettinelli	*La smorfia (1959)*	*Pharmacist*
Domenico Fischietti	*Il mercato di Malmantile (1758)*	*Charlatan*
Pietro Guglielmi	*Il medico burlato (1790)*	*Charlatan*
Wolfgang Amadeus Mozart	*Così fan tutte (1790)*	*Charlatan*
Pietro Generali	*Il medico ciabattino (1819)*	*Charlatan*
Pasquale Sogner	*Quattro prigionieri e un ciarlatano (1832)*	*Charlatan*
Gaetano Donizetti	*L'elisir d'amore (1832)*	*Charlatan*
Luigi and Federico Ricci	*Crispino e la comare or Il medico è la morte (1850)*	*Charlatan*
Geroges Bizet	*Le docteur miracle (1857)*	*Charlatan*
Giuseppe Verdi	*La forza del destino (1862)*	*Surgeon*
Umberto Giordano	*Fedora (1898)*	*Surgeon*
Giovanni Paisiello	*Il Barbiere di Siviglia (1782)*	*Physician/ Charlatan*
Gennaro Astarita	*Il medico parigino or L'ammalato per amore (1791)*	*Physician*
Pasquale Sogner	*Privazione genera desiderio, ossia moglie libera e collo torto (1816)*	*Physician*
Gioachino Rossini	*Il Viaggio a Reims (1825)*	*Physician/ Charlatan*
Giuseppe Malusardi	*L'ammalata e il consulto* (1837)	*Physician*
Gaetano Donizetti	*Don Pasquale* (1843)	*Physician*
		Physician
Jacques Offenbach	*Les contes d'Hoffmann (1881)*	*Physician*
Jules Massenet	*Cendrillon* (1899)	*Physician*

(continued)

Table 2.1 (continued)

Composer	Opera	Health care figures
Claude Debussy	*Pelléas et Mélisande* (1902)	*Physician*
Ermanno Wolf-Ferrari	*L'amore medico* (1913)	*Physician*
Ferruccio Busoni	*Arlecchino (1917)*	*Physician*
Giacomo Puccini	*Gianni Schicchi* (1918)	*Physician*
Pietro Generali	*Il medico e il ciabattino (1918)*	*Physician*
Sergej Prokofiev	*L'amour des trois oranges (1921)*	*Physician*
Alban Berg	*Wozzeck (1923)*	*Physician*
Francis Poulenc	*Dialogue des carmélites (1957)*	*Physician*

doctors find themselves at the sick bed without ever agreeing on the diagnosis and therapy, indeed often arguing over which decision to take.

Doctors, apothecaries, and charlatans, often united in a single not so noble vision of the art of treatment, will remain for decades among the most well-known and "exploited" characters in the lyric field to show the limits of a medicine still unable to act as a scientific discipline capable of addressing and solving health challenges [19].

References

1. Vannoni G. A un dottor della mia sorte. Bologna: Pendragon; 2017.
2. Beghelli M. Pazienti, Medici e Speziali nel Melodramma. Bull Sci Med. 2010;182:9–28.
3. Tanturri A. Il dottore all'opera. Figure di medici e guaritori nel melodramma fra Settecento e Novecento. Aevum. 2019;93:754–5. https://doi.org/10.26350/000193_000056.
4. Riva MA, Bellani L, Turato M, Cesana G. Physicians and alternative medicines in "The Barber of Seville" by Gioachino Rossini: a bicentennial debate. Eur J Int Med. 2015;26:757–8.
5. Heuscher JE. Don Pasquale, Fidelio, and psychiatry. Am J Psychoanal. 1992;52:161–73.
6. Willich SN. Physicians in opera – reflection of medical history and public perception. BMJ. 2006;333:1333–5. https://doi.org/10.1136/bmj.39056.464074.55.
7. Corsini A. Medici ciarlatani e ciarlatani medici. Bologna: Zanichelli; 1922.
8. Cosmacini G. Ciarlataneria e medicina. Cure, maschere, ciarle. Milano: Cortina; 1998.
9. Steptoe A. Mozart, Mesmer and 'Così fan tutte'. Music & Letters. 1986;67:248–55.
10. Stile A. Mesmer et Mozart: un rapport (apparentement) simple. Laboratorio dell'ISPF. 2011;8(1/2):1–12.
11. Lorusso L, Franchini AF, Porro A. Opera and neuroscience. Prog Brain Res. 2015;216:389–409. https://doi.org/10.1016/bs.pbr.2014.11.016.
12. Rhodes Peschel E, Peschel EE. Medicine, music and literature. The figure of Dr. Miracle in Offenbach's Les contes d'Hoffmann. Opera Q. 1985;3:59–71.
13. Marini P. Concerto per farmaco e orchestra. Le tracce dei farmaci nelle composizioni musicali. Torrazza Piemontese: s i e; 2021.
14. Zaugg D. Musik und Pharmazie: Apotheker und Arzneimittel in der Oper. Publications of the Société Suisse d'Histoire de la Pharmacie, vol. 20. Lieberfeld: SGGP/SSHP; 2001.
15. Sironi VA. I farmacisti: commercianti o professionisti? In: Varni A, editor. Storia delle professioni sanitarie in Italia tra Ottocento e Novecento. Bologna: Il Mulino; 2002. p. 77–89.
16. Rocchietta S. Farmacisti ed alchimisti, personaggi di opere liriche, operette e balletti. Minerva Farm. 1960;9:176–81.

17. Rocchietta S. Farmacisti ed alchimisti, personaggi di opere liriche, operette e balletti. Minerva Med. 1960;51:636–42.
18. Rocchietta S. Farmacisti nell'opera lirica Minerva Farmaceutica. 1962;11:40–1.
19. Soriano JB. On doctors and their operas: a critical (and lyrical) analysis of medicine in opera. Chest. 2018;154:409–15. https://doi.org/10.1016/j.chest.2018.03.015.

3 Caring Through Music: Music Therapy and Opera Therapy

Livio Claudio Bressan and Vittorio Alessandro Sironi

3.1 Historical Aspects

The role and contribution of music for treatment and healing was already known and recognized in the ancient and classical world. Reminiscences of this empirical mode of therapy are still found today in many music of the popular world [1], but in classical Greece Apollo, father of Asclepius, considered the protector of medicine and music, deserves to be remembered. Pythagoras, Plato, and Aristotle in some of their writings have highlighted the positive effect of music on the psyche and body, underlining its beneficial relaxing and invigorating effects.

The birth of music therapy as a scientific method of treatment dates back to the first half of the eighteenth century. The first systematic treatise on music therapy was published in 1749. It is the book "Reflections on ancient and modern music with the application to the cure of disease" by the London doctor and musician Richard Brocklesby [2]. This was followed in 1758 by another important contribution by the French doctor Louis Roger, which highlights the importance of music for the treatment of diseases related to a decreased mood [3]. We cannot overlook the importance that Franz Anton Mesmer attributes, in the context of his therapeutic practices, to music as an element for determining hypnotic suggestion. The relations he establishes with Wolfgang Amadeus Mozart in this regard are such that the latter will not hesitate to include more than an explicit reference to mesmerism in his comic opera *Cosi fan tutte* (1790).

L. C. Bressan
Neurology Unit, Ospedale Bassini, ASST-Nord Milano, Cinisello Balsamo, Italy

V. A. Sironi (✉)
History of Medicine, School of Medicine and Surgery, University of Milano-Bicocca, Monza, Italy
e-mail: vittorio.sironi@unimib.it

L. Lorusso et al. (eds.), *Effects of Opera Music from Brain to Body*, Neurocultural Health and Wellbeing,
https://doi.org/10.1007/978-3-031-34769-6_3

The doctor and musician Peter Lichtenthal in his "Treatise on the influence of music on the human body and its use in certain diseases" (1811) [4] is the first to identify which pathological conditions can derive the greatest benefits from music therapy. In his opinion, fevers, both acute and intermittent, chronic pain, asthenic diseases, convulsions, and diseases of the mind (including dementia) are those that have the greatest benefits from music, provided that the doctor takes into account the particular characteristics of every patient in the choice of musical pieces and knows how to integrate this treatment modality with other more traditional therapeutic means.

The scientific positivism of the late nineteenth century and psychoanalysis of the early twentieth century contribute to ensuring that in the early years of the new century the music therapy approach for the treatment of behavioral disorders becomes relevant in the United States thanks to the research carried out by neurologist James Leonard Corning, who he is also the first doctor to experiment, alongside traditional listening to music, the use of lyric pieces taken in particular from works by Richard Wagner [5]. In particular, during the First World War in the Anglo-Saxon world, the music therapy approach was frequently used for the treatment of post-traumatic stress disorder (shellshock) with good clinical results.

In the following decades, music therapy enters more and more frequently among the integrative treatments for the treatment of various pathologies (psychic and neurological in particular, but also rheumatological and oncological); until towards the end of the twentieth century, a series of systematic researches on this modality therapeutic has definitively sanctioned its scientific legitimacy [6].

3.2 The Importance and the Role of Music Therapy

Today there are more and more people throughout the Western world who decide to turn to "complementary" or "unconventional" medicine, to treat chronic ailments such as headaches, back pain, and anxious-depressive neurosis, or in search of hope if affected by diseases incurable or terminal. Without entering into the too passionate controversy between proponents and detractors of these methods, complementary medicines promote human care through the exploration of every aspect: physical, psychological, emotional, cognitive, social, and spiritual. One of the most frequent criticisms of the proponents of nonconventional medicines to contemporary medical practice is the inattention to the emotional side of the disease, therefore the excessive effort in demonstrating "the objectivity of the observed data." According to this opinion, the excessive search for "objectivity" of scientific medicine would end up diverting the act of health care from its original purposes. On the contrary, the disease, according to a modern "holistic" point of view, should be faced with integrated assistance, aimed at improving the quality of life, where health procedures, priority for the biological aspects of the therapy, can be accompanied by other nonmedical interventions, but always aimed at achieving the psychophysical well-being of the person. Recent data from the World Health Organization (WHO) indicate that 15% of Italians resort to complementary medicine at least once a year.

In the United States and in the rest of Europe, "holistic" interventions are more widespread and are usually conducted in full autonomy by an operator expert in complementary methodologies, in agreement with doctors, nurses, and other health professionals.

Music therapy is currently one of the most widely used complementary methodologies in the health and social care sectors. According to the musicologist Pier Luigi Postacchini, it can be defined as "a technique by which various professional figures facilitate the implementation of projects of spatial, temporal and social integration of the individual, through strategies of harmonization of the functional structure of the handicap, through the use of the musical parameter [...], pursued with a work of affective tuning, which are possible and facilitated thanks to specific strategies of non-verbal communication" [7]. This vision is today highlighted according to the most current and specific definition of the World Federation of Music Therapy (WFMT), for which "music therapy is the use of music and/or musical elements (sound, rhythm, melody and harmony) by of a qualified music therapist, with a client or a group, in a process aimed at facilitating and promoting communication, relationships, learning, motor skills, expression, organization and other relevant therapeutic objectives, in order to satisfy his physical, emotional, mental, social and cognitive needs" [8]. In other words, music therapy aims to develop the potential and/or residual functions of the individual in such a way that it can better achieve intra- and interpersonal and, consequently, can improve the quality of one's life thanks to a preventive, rehabilitative, or therapeutic process.

Music therapy has precise neuroscientific bases that start from the principle that sensory input is able to facilitate, through the structures of the temporal lobe, the processes of attention, observation, and learning managed by the frontal lobe; it can be hypothesized that pleasure is conscious or unconscious, evoked by listening and/or by musical practice, that is able to trigger neurobiological processes of a reparative type. In other words, the harmony, melody, and rhythm of a piece of music, enjoyed in a context of harmony and empathy with the group and the therapist, could lead back conflicting parts of the mind-body to a cognitive and affective-relational recompositing. That is, music, with its facilitating pleasure and sometimes of reparative regression, is able to penetrate the circuits of the central nervous system facilitating thought and rebalancing behavior.

There is now considerable scientific evidence regarding the mechanisms underlying the functioning of music therapy. In the last twenty years, functional neuroimaging of the central nervous system has opened unimaginable scenarios until a few decades ago, on the neurobiological involvement of anatomical structures responsible for learning, thinking, and behavior following musical stimulations. As far as our experience is concerned, neurological diseases of a chronic and progressive degenerative type (such as Parkinson's disease and Alzheimer's disease) have greatly benefited from music therapy treatments administered with standardized techniques, with statistically significant benefits, detected with scales and validated, both functional and in terms of quality of life.

3.3 Clinical Applications of Music Therapy

There are many pathologies that can benefit from music therapy. The music therapist has the ability to work with a variety of patients, both children and adults, who may have emotional, physical, mental, or psychological handicaps. Therefore, music therapy can be applied to all age groups and in a variety of care settings. Music, in fact, has a nonverbal quality and offers a wide range of verbal and vocal expression. The sonorous-musical element of music can acquire therapeutic connotations in two operational sectors: psychotherapy and rehabilitation. In the psychotherapy sector, the sound-musical element becomes a nonverbal communication channel that can favor the establishment of particular forms of interpersonal relationships. In the rehabilitation sector, on the other hand, music acts as a stimulus for a specific function (motor, vocal, cognitive, etc.) and constitutes the formal model within which this function is articulated [9].

An intense research work has been underway for some decades, with participation in conferences and congresses and through scientific publications in national and international journals, with the ultimate aim of contributing to better "characterize" this discipline and then "clear it through customs," once validated, in the field of modern neuroscience. However, there is still a long way to go and an increasingly intense comparison of clinical research is required, a further discussion of both paradigmatic and unusual clinical cases, and the preparation of multiple research protocols aimed at the acquisition of further scientific data.

How do you choose a "therapeutic" piece of music? It must ignore aesthetic considerations (such as "beautiful and pleasant music"), but the musical piece must rather have the purpose of stimulating dormant psychic areas, involving more complex affective levels. Some music, more than others, have these powers and are the ones that affect the world of affection. In terms of musical choices, there is a somewhat confused literature because often the author and the genre of music are decided without taking into account the "type of listener." On the contrary, we believe that the patient should be at the center of our interests; otherwise, the risk is that of offering music that only appeals to the one who administers it. Only a careful musical history will allow us to understand what musical culture the patient possesses and in what sound environment he has lived.

But what are the peculiar characteristics that give music a therapeutic efficacy? We will try to propose some possible answers. New music, that is, never listened to, leads to the development of some fundamental psychic functions such as attention and a certain expectation which, in turn, are able to arouse emotions and new associations. Listening to music never heard before can be "kidnapped" and induced to explore a new planet. However, we believe that even the musical pieces already known—that is, whose development of musical themes we know—by evoking memories linked to specific temporal and spatial circumstances can make us achieve interesting goals. Furthermore, most scholars of music therapy seem to favor classical music, attributing only therapeutic powers to it. There is no doubt that there has been greater research in this musical area, probably because in addition to proposing a rich and articulated message, this is also a musical genre that has already been widely tested over time and known to the general public.

For our therapeutic purposes we have built different sound paths, which are adapted from time to time to the individual patient or to the group and which include pieces of classical music, jazz, pop, light music, from films, etc. However, we want to affirm with conviction that the secret of the *therapeutic path* lies not so much in the choice of the type of musical piece, as in the logical placement of the pieces themselves. That is, the problem is the following: once the starting piece has been identified, which other pieces should we add and in what order to obtain the result we want? There is no precise rule. And this is why an accurate musical anamnesis is required, carried out by a music therapist with a high technical and cultural profile.

Music therapy is indicated in different types of pathologies, since even patients who are not cooperating due to their physical conditions (bedridden, disabled, or in a coma), or psychic (with dementia or mental retardation), or very young children, can benefit from it. In particular, a great efficacy is universally evident particularly in four specific sectors: palliative care, vegetative states, Alzheimer's dementia, and Parkinson's disease [10].

Palliative Care Music is used in clinical situations of chronic pain, in terminal pathologies, and in their therapies. Intractable pain is one of the most frequent and painful consequences of neoplastic pathologies and, often, it is necessary to resort to narcotics that are not free from side effects. In this context, receptive music therapy has been used successfully in relieving neoplastic pain in patients receiving fixed doses of analgesics. On the basis of these observations and a long French tradition dating back to the monks of Cluny (eleventh century), music thanatology was proposed, intended as a palliative medical intervention used to satisfy the complex physical and spiritual needs of the dying patient. In other words, music can also be used as a means to improve the patient's "quality of death," when there is nothing left to do but accompany him "lovingly" in the last steps of his journey.

Vegetative States The use of music therapy in the state of coma is current and of great prominence. Sound stimulation, especially if relevant to the patient's emotional memory, can play a role in the attempt to establish contact, acting on an affective, cognitive, and relational level.

Dementias Among the neurogeriatric pathologies, music therapy finds an important implementation in dementias, especially in relation to the characteristic behavior and mood disorders. In fact, music can be used to reduce aggression, especially during meals, or to improve the quality of life of institutionalized patients. In Alzheimer's disease, the music therapy intervention was also aimed at the rehabilitation of specific motor, cognitive, and psychic functions [11].

Parkinson's Disease In this pathology a method of active music therapy has recently been proposed for the recovery of emotional and motor functions, characterized by the loss of rhythm and kinetic synchrony. This method has been articulated in such a way as to use musical instruments, the body, and the voice in making music, or even "noise," to stimulate the recovery of harmony and speed of movement, the vocal tone, and the expressiveness of the face with improvement of the quality of life. Focusing on the physical, psycho-emotional, and social impact of Parkinson's disease from the point of view of the sick individual, a coordinated set of rehabilitation interventions was developed, assigning priorities according to the repercussions of these interventions on the quality of life of the patient. These rehabilitation interventions—which are based on a harmonious integration between scientific medicine and complementary medicine—have the ultimate aim of alleviating the suffering of Parkinson's patients to whom it is unfair to promise an unattainable recovery, but with which the necessary synergies must be made, to avoid isolation and to combat that sense of helplessness that takes both the sick and their families [12–17].

It can therefore be said that the most significant qualitative leap that has taken place in recent years in the therapy of a disease is to be ascribed to the awareness, ever more explicit and widespread, of the value of a new and different *philosophy of management of the sick person*, at the same time global and individualized, capable of placing at the center of its interests, not only motor disorders, but the "person" as a whole. The consequences deriving from this change of perspective are many and profound: the primary target of the treatment becomes the *patient's quality of life* and not just physical autonomy. Not only that, the treatment is transformed from a simple administration of active ingredients into an *integrated and personalized treatment plan* and the information of the patient and his family members has the aim of reaching not only a better *compliance,* but a real *therapeutic alliance*. For this purpose, the use of music therapy fits well, which in the sense of the term is placed among the complementary bio-natural disciplines that aim to maintain or recover the person's state of well-being. In fact, music therapy is deliberately not intended to "replace" official medicine, but to integrate with it, stimulating the person's vital resources through natural facilities such as the sound, rhythm, and noises of nature.

3.4 A New Dimension of Care: Opera Therapy

Historically, after the aforementioned initial sporadic attempts to use James Leonard Corning's listening to lyrical music for curative purposes, also in the United States, during the years of the Great Depression, a large program called the Federal Music Project promoted by the government and led by the musician Nikolai Sokoloff, it was aimed in particular at people of different ages with neuropsychic disorders (Fig. 3.1).

It used different musical genres for rehabilitation and therapeutic purposes, including melodrama (opera and operetta), drawing a wide range of sensations from their music: contentment/sadness, happiness/melancholy, pleasure/pain, and restlessness/vivacity [18]. After these first experiences of opera therapy only in the last two decades have systematic scientific research been undertaken to demonstrate the positive influence of this integrative music therapy approach in medicine [19]. The role of classical music in the treatment of numerous pathologies has been known for some time [20, 21], while there are few studies on the possible benefits of melodrama in the medical field. The study of the variation of physiological parameters in healthy subjects (professional musicians or simple passionate music listeners) highlighted in the monitored subjects significant correlations between cardiovascular picture (decrease and regularization of heart rhythm and blood pressure), respiratory rate, and musical profile, in particular with reference to two of the most famous operas by Italian composers—such as some arias from Giacomo Puccini's *Turandot* (1926) and Giuseppe Verdi's "Va, Pensiero" choir of *Nabucco* (1842)—with no qualitative difference between musicians and nonmusicians [22] (Fig. 3.2).

Fig. 3.1 Federal Music Project. Federal Music Project, U.S. *Works Progress Administration Federal Music Project of New York City Theatre of Music Presenting chamber operas, symphony concerts, grand operas, and chamber music at popular prices*. New York, None. [Nyc: federal art project, between 1936 and 1941] [Photograph] Retrieved from the Library of Congress. Laquatra, J., Mérimée, P. and Federal Music Project, U.S. (1939) *"Carmen" Presented by Cuyahoga County Opera Association and the Federal Music Project: Ballet directed by Madame Bianca*. Ohio, 1939. Ohio: Federal Art Project. [Photograph] Retrieved from the Library of Congress

These results have helped to understand how opera music can transmit emotions capable of improving specific cardiovascular responses, both in healthy subjects, a function of a better increase in some sports performances, and in sick subjects such as heart patients. For these effects mediated by the autonomic nervous system, some studies have shown how the auditory and visual use of melodrama improves the performance of students in schools and universities.

Opera can also play a positive role in overcoming social differences and ethnic discrimination between people of different cultural backgrounds, managing to increase the degree of tolerance and mutual understanding. On the basis of these encouraging results, studies have been undertaken to evaluate the use of the practice of operotherapy in various diseases such as cancer, disturbances of consciousness, and coma, opening new dimensions of integrated care in the context of frequent and recurrent pathologies [23].

Studies with functional magnetic resonance (fMR) show that, although there are no real "musical centers" in the brain, however, compared to listening to a generic noise with a fixed tone, the brain areas activated by music, even if adjacent, are different from those involved in language (both in listening and in reading). This anatomical-functional separation confirms the great autonomy of music with respect to language, regardless of whether the code is visual or auditory. Neuroscience therefore confirms that even simple listening to music is a formidable way to put into action or reactivate multiple neuronal networks: from attention to memory, from learning to creativity, and from emotional to motor processes.

They also show how the brain is able to select and understand the emotional nuances of sounds through specialized neuronal populations of the frontotemporal cortex, both of the voice and of music, distinguishing the connotations that evoke positive sensations from those that provoke negative sensations: a vocal/musical intertwining that blends wonderfully in operas.

Music then, especially when it interacts with singing and movement (as happens in opera and ballet), is a formidable tool capable of positively affecting neuroplasticity, which is the ability to "remodel" synaptic connections (the physical modalities through which neurons speak and communicate with each other) and a formidable generator of neuromodulators (the liquid words) positive for our well-being: serotonin, dopamine, and endorphins. This is why music can represent, as we have seen, an excellent therapy for serious diseases: from Parkinson's to Alzheimer's, from depression to acute anxiety. And next to music therapy even more so is opera therapy, that is, listening to and viewing the opera, which, by combining music, singing, and gestures, plays an even more beneficial action on the healthy and more distinctly therapeutic on the sick [24, 25].

The importance of a possible chromatic component in listening to music had already been noted and highlighted in the past (Lichtental mentions it in his work on the influence of music on the human body, cited above), but a scientific understanding of this experience is today possible thanks to that sensorial/perceptive neurological phenomenon called synesthesia, which is a "contamination" of the senses in the perception that occurs when stimuli that arrive through a sensory or cognitive path induce automatic experiences related to another sensory or cognitive path: for

Va, pensiero, sull'ali dorate;
va, ti posa sui clivi, sui colli,
ove olezzano tepide e molli
l'aure dolci del suolo natal!

Fly, my thoughts, on wings of gold;
go settle upon the slopes and the hills,
where, soft and mild, the sweet airs
of my native land smell fragrant!

Fig. 3.2 The first passages of "Va, pensiero" *Nabucco* (1842) of Giuseppe Verdi. Promotional poster, by Illustrator Leopoldo Metlicovitz (1868–1944) for Giacomo Puccini's opera "Turandot," on April 25, 1926

example, in our case "seeing" the musical notes as colors and observing a chromatic overview while listening to music is a not infrequent synaesthesis experience [26–28]. Paradoxically, the association of colors and sounds can be better influenced and modulated during the viewing and listening of the opera, which can already materially show during the performance of a piece of music, the colors of the costumes, the chromatism of the scenography, or the color dynamics of ballets.

3.5 Conclusion

Compared to traditional music therapy, opera therapy determines a greater and more intense emotional, sensory, and physical involvement. Even listening to music and singing alone play a more important role of "neuropsychological entrainment" than traditional music therapy. Attending an opera performance—both in the theater and even simply through the recording of the melodrama—determines an even more powerful involvement. It implies the use of sight (which is stimulated by the scenography and stage costumes) and motor participation (determined by the movement of the characters on stage and by the development of the narrative action) which probably involve the "mirror neurons" in such a way to provoke a psycho-physical "rebalancing" capable of positively influencing multiple pathological conditions.

In this sense, it is conceivable to assume that changes in brain neuroplasticity may over time be induced both by music therapy and even more by opera therapy and that these brain changes are at the origin of the observed therapeutic benefits. This perspective suggests that the well-established use of music therapy and the use of opera therapy still in the experimental phase may represent real and important integrative therapeutic possibilities to improve the treatment of the suffering of sick patients.

References

1. Sironi VA, Riva MA. Neurological implication and neuropsychological consideration on folk music and dance. Prog Brain Res. 2015;217:187–205.
2. Brocklesby R. Reflections on ancient and modern music with the application to the cure of disease. London: Cooper; 1749.
3. Roger L. Traités des effects de la musique sur le corp humain. Paris: Brunot; 1801.
4. Lichtenthal P. Trattato dell'influenza della musica sul corpo umano e sul suo uso in certe malattie. Milano: Silvestri; 1811.
5. Davis WB. The first systematic experimentation in music therapy: the genius of James Leonard corning. J Music Ther. 2012;49:102–17.
6. Rose FC, editor. Neurology of music. London: Imperial College Press; 2010.
7. Postacchini PL, Ricciotti A, Borghesi M. Musicoterapia. Roma: Carocci; 1997.
8. Bunt L, Hoskyns S, Swami S. The handbook of music therapy. London: Routledge; 2013.
9. Altenmüller E, Schlaug G. Neurologic music therapy: the beneficial effects of music making on neurorehabilitation. Acoust Sci Technol. 2013;34:5–12.

10. Brancatisano O, Baird A, Thompson WF. Why is music therapeutic for neurological disorders? The therapeutic music capacities model. Neurosci Biobehav Rev. 2020;112:600–15.
11. Dowson B, Schneider J. Online singing groups for people with dementia: scoping review. Public Health. 2021;194:196–201.
12. Masahiro O, Masahiro S, Kazutoshi K. Paired synchronous rhythmic finger tapping without an external timing cue shows greater speed increases relative to those for solo tapping. Sci Rep. 2017;7(1):43987. https://doi.org/10.1038/srep43987.
13. Thaut MH, Ruth RR, Thenille BJ, Corene PH, McIntosh GC. Rhythmicauditory stimulation for reduction of falls in Parkinson's disease: a randomized controlled study. Clin Rehabil. 2019;33(1):34–43.
14. Devlin K, Alshaikh JT, Pantelyat A. Music therapy and music-based interventions for movement disorders. Curr Neurol Neurosci Rep. 2019;19:83.
15. Barnish MS, Barran SM. A systematic review of active group-based dance, singing, music therapy and theatrical interventions for quality of life, functional communication, speech, motor function and cognitive status in people with Parkinson's disease. BMC Neurol. 2020;20:371.
16. Bergna A, Zanfagna E, Ballabio A, Vendramini A, Bressan LC, Solimene U. Efficacy of the osteopathic treatment in Parkinson's disease. Bull Rehabilitat Med (Вестник восстановительной медицины). 2021;20(2).
17. Fodor DM, Breda XM, Valean D, Marta MM, Perju-Dumbrava L. Music as add on therapy in the rehabilitation program of Parkinson's disease patients: a Romanian pilot study. Brain Sci. 2021;11:569.
18. Davis WB. Music therapy practice in new York City: a report from a panel of experts. J Music Ther. 1997;34:68–81.
19. Lorusso L, Porro A. Opera and neuroscience: a historical approach and its relevance today. In: Colombo B, editor. Brain and art. Cham: Springer Nature; 2020. p. 81–94.
20. Edwards J. The use of music in healthcare contexts. A select review of writings from 1890s to the 1940s. Voice. 2008(2):1–19.
21. Gardner B. Therapeutic qualities of music. Mus Lett. 1944;25:181–6.
22. Bernardi L, Porta C, Casucci G, et al. Dynamic interactions between musical, cardiovascular and cerebral rhythms in humans. Circulation. 2009;119:3171–80.
23. Young H, Baker F, Shoemark H. Song preference of Chinese older adults living in Australia. Aust J Mus Ther. 2014;25:103–21.
24. Altea P, Bellavigna R, Camaghi R, Corti N, Delicati F, Downie G, Pozzi L, Corno L. Musicoterapia con il malato di Alzheimer, Progetto Anziani Musicoterapia PAM. Milano: Federazione Alzheimer Italia; 2003.
25. Leggieri M, Thaut MH, Fornazzari L, Schweizer TA, Barfett J, Munoz DG, Fischer CE. Music intervention approaches for Alzheimer's disease: a review of the literature. Front Neurosci. 2019;13 https://doi.org/10.3389/fnins.2019.00132.
26. Braganca GFF, Fonseca JGM, Caramelli P. Synesthesia and music perception. Dement. Neuropsychology. 2015;9:16–23.
27. Triarhou LC. Neuromusicology of Musiconeurology? "Omni-art" in Alexander Scriabin as a Fount of ideas. Front Psychol. 2018;7:1–7. https://doi.org/10.3389/psyg.2016,00364.
28. Lorusso L, Ione A, Franchini FA, Porro A. Music synesthesia and wellbeing. In: Colombo B, editor. The musical neurons. Cham: Springer Nature; 2022.

Part II

ACT II Relationship Between Music and the Brain

4 "Va Sossopra Il Mio Cervello" (My Brain Goes Up): Neurology and Psychiatry in the Opera

Alessandro Porro, Francesco Brigo, and Lorenzo Lorusso

4.1 Introduction

Although from a historical point of view the relationship between music and medicine is evident [1], only recently have historians of medicine and music shown interest in the description of medical aspects in operas [2–6]. As with any artistic form, the representation of neuropsychiatric disorders in operas reflects the cultural and social attitudes towards mental illnesses and neurological conditions in various historical periods [4, 7].

The presence of neurological [8] and psychiatric disorders in operas, either implicit or explicit, cannot be fully understood without referring to the medical theories of the time. Since the operatic genre dates back over 500 years, the ancient theory of humors should be briefly mentioned, as it remained valid until the entire mid-nineteenth century. According to it, neuropsychiatric disorders are strictly related to the melancholic temperament, characterized by a preponderance of black bile over the other three humors (blood, phlegm, and yellow bile). The English language term "melancholic" is derived from the Greek term meaning "black bile"; similar terms can be found in many Romance languages, such as French, Castilian, or Portuguese, while in Italian and Catalan, it underwent a metathesis leading to the terms "malinconico" and "malenconic."

A. Porro
History of Medicine, University of Milan, Milan, Italy

F. Brigo
Department of Neurology, Hospital of Merano-Meran (SABES-ASDAA), Merano-Meran, Italy

L. Lorusso (✉)
Neurology Unit, ASST-Lecco, Merate Hospital, Merate, Italy
e-mail: l.lorusso@asst-lecco.it

L. Lorusso et al. (eds.), *Effects of Opera Music from Brain to Body*, Neurocultural Health and Wellbeing,
https://doi.org/10.1007/978-3-031-34769-6_4

The concept of melancholic temperament can, therefore, serve as a guide in considering how neuropsychiatric disorders and madness were represented over time in operas, even if the humoral theory was never explicitly mentioned in musical theatre. However, in antiquity, temperaments were also defined as "affective complexions," making evident the connection with emotions (or "affects") expressed through opera. Even the couples of antithetical qualities that intrinsically characterize the four humors find their correspondence in the couples of "affects" depicted in the music. This emphasizes the deep relationship between the "theory of affects" and the theory of intervals and tones [9, 10].

Interestingly, traces of this ancient concept of "theory of affects" can be found throughout the entire century-long history of music, until its present days. As an example, the symphony No. 2 by Carl Nielsen (1865–1931) carries the title "The four temperaments," reflecting a sort of neo-hippocratism deeply imbued with psychoanalysis. It is perhaps no coincidence that in the twentieth century, psychoanalysis has emerged as a new "theory of affects," representing an interpretative paradigm for the analysis of operas [11].

Since the early seventeenth century, madness and other neuropsychiatric disorders have been represented in operas in ways that reflect the cultural and social attitudes towards mental illnesses and insanity in various historical periods.

4.2 At the Beginning of Musical Theater

Alongside melancholia, at the beginning of musical theater, the disorders that today we would regard as neuropsychiatric were perceived and represented according to the concept of *passiones animi* ("passions of the soul"). These passions had to be controlled through different means, for instance, through behavior aimed at maintaining health and preventing disease, as diseases were thought to originate from these *passiones animi* becoming out of control.

The case of falling in love was exemplary and paradigmatic of these uncontrolled "passions of the soul," especially when affecting an elderly male person. Theatrically, it was further emphasized by the social divide: in the comic opera, the dualism between the rich old man and the smart young woman was frequently found in operatic plot, with the intertwining of situations and the inevitable happy ending (consider, for instance, *Don Pasquale* by Gaetano Donizetti and *Falstaff* by Giuseppe Verdi). The plot did not always result in the fulfillment of the aspirations of the elderly, but good often triumphed, thanks also to the artifice of sublimation, with the overcoming and rebalancing of the "passions of the soul." Of note, this general schema of the opera plot suited particularly well with the inclusion of the concepts and situations of the Commedia dell'Arte in the musical theatre. Female love madness was also represented in operas, particularly in the nineteenth century, reflecting the evolution of medical nosography and the appearance of disorders believed to be typically feminine, such as hysteria.

Besides emotional and affective derangement, a further condition often represented in operas is that of mental derangement and madness, a theme that is often derived from literary sources. However, we should be aware that, in the past, madness was conceived and interpreted according to nosological theories that are no longer valid or grouped into nosographic entities that nowadays appear to be completely distinct. This implies that, in an inferential dimension, certain traditional identifications can or must be questioned. How many operatic representations of a mental derangement that were generically termed "madness" should not, instead, be attributed to dementia?

However, to analyze the representation of madness in early musical theater, one should consider other aspects related to the social climate of the time. Madness was generally conceived as a derangement from the normal state and was interpreted not only as an attitude or an effect but also as a cause: anyone who deviated from the norm was ipso facto framed into the vague realm of madness.

This condition implied not only separation but also expulsion from the social assembly. However, in the past, marginalization and expulsion did not always require the presence of a physical disorder (consider, for example, the anti-Semitic features of some Italian operas from the sixteenth century, such as the *Amfiparnaso* by Orazio Vecchi [1550–1605]) [12] (Fig. 4.1). Tomaso Garzoni (1549–1589) in his opera entitled *La piazza vniversale di tvtte le professioni del mondo, e nobili et ignobili* [13] presents a series of dishonorable professions, including also the itinerant theater led by converted Jews (Fig. 4.2).

The following year he composed the opera *L'Hospidale de' pazzi incurabili* [14], in which he dealt with the conventional topoi of madness and melancholy.

As for the indirect presence of madness in the context of musical theater, we can also recall the relation between comedy and the so-called *intermedii*. The Florentine *intermedii* were represented in 1589 and composed on the occasion of the arrival in Florence of Christine of Lorraine (1565–1636), who had married Ferdinando I de 'Medici three years earlier (1549–1609). On May 13, 1589, they were represented with the comedy *La Pazzia d'Isabella*. It was represented by the *Compagnia dei*

Fig. 4.1 Engraved scene from the opera Amfiparnaso, Venice, 1597. Portrait of Orazio Vecchi (1550–1605)

Fig. 4.2 La piazza vniversale di tvtte le professioni del mondo, e nobili et ignobili by Tomaso Garzoni (1549–1589)

Comici gelosi, led by Francesco Andreini (1548–1624) and by his wife Isabella Canali (1562–1604), who was the author and protagonist of the play herself.

Towards the end of the nineteenth century, the representation of madness progressively assumed a more modern connotation and began being regarded as a true neuropsychiatric, pathological disorder (consider, for instance, Rossini's representations, in which the reason derails in an almost mathematical way [15]). This occurred in parallel with the development of modern neuropsychiatry. Thanks to the contributions of Jean Etienne Dominique Esquirol (1772–1840), Philippe Pinel (1745–1826), and Jean-Martin Charcot (1825–1893), neuropsychiatry profoundly influenced the neuroscientific progress of the second half of the nineteenth century, leading to several studies on neuroses and the emergence of new nosographic entities [4].

In nineteenth-century operas, it is not always easy or possible to differentiate between madness and hysteria (a condition that is attested to in several operas of this period [16]. Hysteria can already be glimpsed in Berlioz's Symphonie fantastique (1830), with the protagonist being obsessed by the idée fixe of his love for his lover. The character of Lucia in Donizetti's *Lucia di Lammermoor* [17] was considered almost an emblem of hysteria: it is surely no coincidence that Emma Bovary (whom Baudelaire considered suffering from hysteria in an article published on October 18, 1857, in L'Artiste) was so fascinated by this character. In *Tristan und Isolde*, Wagner staged the all-psychic discomfort of the protagonists, representing their "neurosis" and emphasizing what Freud would have called "compulsion to repeat" [18]. Friedrich Nietzsche, in his ferocious attack against Wagner (Nietzsche contra Wagner), defined the composer as "une névrose," drawing a parallel – never explicit but extremely obvious, especially for his contemporaries – with Charcot and hysteria.

4.3 A Reasoned List of Operas

Herein, we propose a reasoned list of operas in which the aspects relating to neurological and psychiatric pathologies are prominent or can be easily recognized.

The first group of operas is centered around the mythological theme of madness. They include *Il ritorno di Ulisse in Patria* (1640) by Claudio Monteverdi (1567–1643), *Mèdée* (1693) by Marc-Antoine Charpentier (1643–1704), *Dido* (1641) and *L'Egisto* (1643) by Francesco Cavalli (1602–1676), *Totila* (1677) by Giovanni Legrenzi (1626–1690), *Atys* (1676) and *Roland* (1685) by Jean-Baptiste Lully (1632–1687), and *Orlando generoso* (1691) by Agostino Steffani (1654–1728). The theme of madness simulation appears in *La finta pazza Licori* (*The feigned madwoman Licori*) (1627), an unfinished and lost opera by Claudio Monteverdi, later reprised by Francesco Sacrati (1605–1650) with the title *La finta pazza* (1641), both operas on a libretto by Giulio Strozzi (1583–1652). The mythological theme of madness was also the subject of eighteenth-century operas: *Orlando furioso* (1713) and *Orlando finto pazzo* (1714) by Antonio Vivaldi (1678–1741), *Orlando* (1733) by George Frideric Haendel (1685–1759), and *Orlando paladin* (1782) by Franz Joseph Haydn (1732–1809), with *Iphigénie en Tauride* (1779) by Christoph Willibald Gluck (1714–1787) and *Roland* (1778) by the Italian Niccolò Piccinni (1728–1800).

The problems of lunatic asylums were shown in a comic opera by Baldassare Galuppi (1706–1793) on a text by Carlo Goldoni (1707–1793) entitled: *Arcifanfano—Re dei matti* (*Arcifanfano—King of fools*) (1749) [19]. In this period, following sociocultural changes, madness is charged with a new meaning: it combines the social fear of the medical condition with its moral implications. The disorder that affects the soul of the mad is due to exaggerated passions that lead to committing sins and manifests itself as a punishment for such excesses [20].

Towards the end of the eighteenth century, the Enlightenment promoted new approaches to madness [21]. The operas of this period were influenced by the classicism associated with a romantic atmosphere, as in *The Ephesian Matron* (1769) by the Englishman Charles Dibdin (1745–1814), *La finta giardiniera* by Wolfgang Amadeus Mozart (1756–1791), and *Nina, o sia La pazza per amore* (1786) by Giovanni Paisiello (1740–1816). As already mentioned, these operas shared some characters and situations with those of the Commedia dell'arte. In general, they represented heroines with a mind upset by the adversities of fate, and they allowed the singer to demonstrate her virtuosity, especially if delirium was represented on stage.

In the nineteenth century, the category of "*folles par amour*" became more established as scenes depicting madness began to be increasingly entrusted to female protagonists who had lost their minds after a disappointment in love [22, 23].

There are rare examples of madness due to remorse of conscience resulting from criminal acts that are the prerogative of men, as in *Semiramide* (1823) by Gioachino Rossini (1792–1868), *L'esule di Roma* (1828) by Gaetano Donizetti (1797–1848), and *I Masnadieri* (1847) by Giuseppe Verdi (1813–1901). Verdi represented delirium in *Nabucco* (1842), *Don Carlo* (1867), and *La Traviata* (1853) [24]. Male madness was also represented by the Russian composer Modest Petrovič Musorgsky (1839–1881) in *Boris Godunov* (1869, revised 1872). Another form of madness is that caused by external agents that cloud the mind and are at the basis of delusional manifestations, as in *Masaniello, La Muette de Portici* (1828) by Daniel Auber (1782–1871) or *Il Pirata* (1827) by Vincenzo Bellini (1801–1835). Finally, there are simulated madnesses, as in *I pazzi per progetto* (1830) by Donizetti or *Italiana in Algeri* (1813) by Gioachino Rossini. As aforementioned, amorous madness was represented in nineteenth-century operas with a stereotyped scheme, the so-called scene of madness, which was well codified and recognizable by the contemporary public of the time. Examples of operas including a scene of madness are *Anna Bolena* (1830), *Gemma di Vergy* (1834), *Lucia di Lammermoor* (1835), and *Linda di Chamounix* (1842) by Donizetti. This Italian composer represented male madness in Maria Padilla (1841), Torquato Tasso (1833), and *Il furioso all'isola di San Domingo* (1833), where a character is affected by a conversion disorder (hysterical blindness). Other operas presenting female madness are *I puritani* (1835) and *Il Pirata* by Bellini, the *Étoile du nord* (1854) by Giacomo Meyerbeer (1791–1864), *Ruddigore* or the *Witch's Curse* (1887) by Arthur Sullivan (1842–1900), and many others [4, 7, 16, 20, 21] (Fig. 4.3).

Starting in the nineteenth century, opera was influenced by emerging neuropsychological theories and began representing patients suffering from neurological pathologies such as sleepwalking; see Bellini's *Sonnambula* (1831) and Verdi's *Macbeth* (original version 1847, revised 1865). Other composers dealing with the theme of sleepwalking were Luigi Alessandro or Lodovico Piccinni (1779–1850), son of the more famous Niccolò, author of a *Sonnambula* (1796), Ferdinando Paër (1771–1839) with *Sonnambula* (1800), the Berliner Karl Blum (1786–1844) with *Die Nachtwandlerin* (1822), Sicilian Salvatore Agnelli (1817–1874) with *Il fantasma* (1842), and Giuseppe Persiani (1799–1869) with *Il fantasma* (1843). Episodes of male sleepwalking are represented in *Il matrimonio segreto* (1791) by Domenico Cimarosa (1749–1801), *Il sonnambulo* (1824) by Michele Carafa (1787–1872), *Der Vampyr* (1828) by Heinrich August Marschner (1795–1861), *Il sonnambulo* (1829) by Luigi Ricci (1805 1859), *Il sonnambulo* (1834) by Carlo Valentini (1790–1853), *O sonámbulo* (1835) by Luis António Miró (1815–1853), and *Der Prinz von Homburg* (1935) by the German composers Paul Graener (1872–1944) and Hans Werner Henze (1926–2012) (1958, new version, 1991) [5, 25–27]. A character with migraine, somnambulism, and rapid eye movement (REM) behavior disorder was depicted in *Il Matrimonio Segreto* (1792) by Domenico Cimarosa (1749–1801) [28].

Epilepsy was also represented in some operas: *Ivan IV* (1862–1865, first performance 1946) by Georges Bizet (1838–1875), *Otello* (1887) by Giuseppe Verdi, and possibly *Pagliacci* (1892) by Ruggero Leoncavallo (1857–1919) [4, 5]. Migraine is present in the aforementioned Cimarosa's *The Secret Marriage*. Generic headache

Fig. 4.3 Poster for the production of *Ruddigore*, 1887. William Schwench Gilbert (1836–1911) and Arthur Sullivan (1842–1900), composers

is represented in Verdi's *Otello,* and in *Love for three oranges* (*Ljubov k trëm apel'sinam*) by Sergej Sergeevič Prokof'ev (1891–1953) and possibly in *Siegfried* (1876) by Richard Wagner (1813–1883) [29, 30] (Fig. 4.4).

The transition to the twentieth century was characterized by the development of psychoanalysis and modern psychiatry and neurology. The influence of the theories of Sigmund Freud (1856–1939) is well evident in *Salome* (1905) and *Elektra* (1909) by Richard Strauss (1864–1949), *Die Gezeichneten* (The Branded or The Stigmatized) (1918) by Franz Schreker (1878–1934), and characters with personality disorders in the operas *Kát'a Kabanová* (1921) and *Osud* (1907/1934) by Leoš Janáček (1854–1928), with *Erwartung* (1924) by Arnold Schoenberg (1874–1951) and *La cena delle beffe* (1924) by Umberto Giordano (1867–1948). In *Wozzeck*

Fig. 4.4 Drawing by Peter Hoffer (1924–2000) for the libretto of *Il Matrimonio segreto*, 1954 (Archivio Ricordi). Domenico Cimarosa (1749–1801), 1870s painting by Francesco Saverio Candido (1768–1807), Museo di San Martino (Naples)

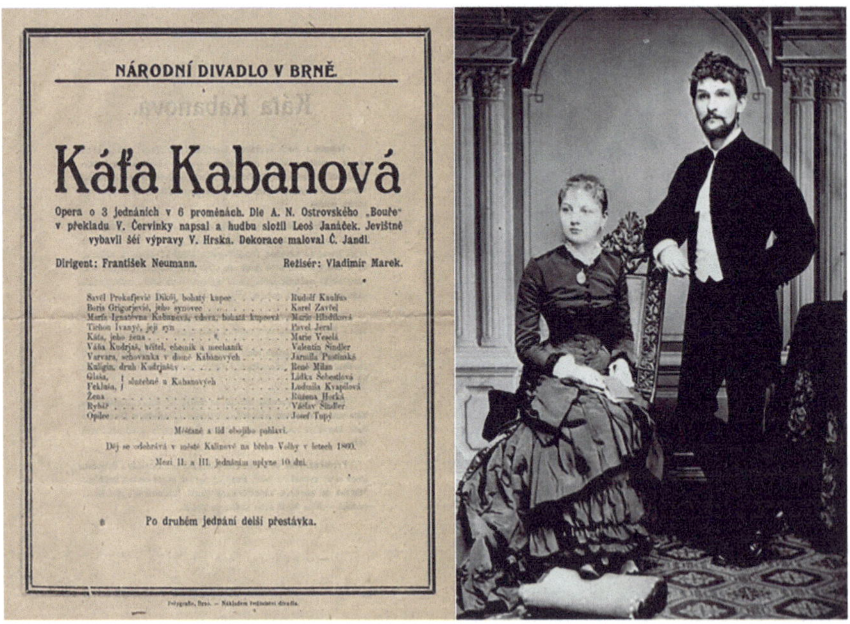

Fig. 4.5 Front page of *Kát'a Kabanová* (1921) by Leoš Janáček (1854–1928). The Czech Leoš Janáček and his wife in 1881, archive of the Department of Music History of the Moravian Regional Museum in Brno

(1925), by the Austrian Alban Berg (1885–1935), alongside psychiatric problems, there is also a description of the risk factors for stroke (Fig. 4.5).

More recent operas have presented a more well-defined clinical connotation based on the progress of medical-scientific knowledge in the fields of psychiatry and neurology: *The Rake's Progress* (1951) by Igor Stravinsky (1882–1971) shows a form of insanity secondary to tertiary syphilis. *Peter Grimes* (1945), *The rape of Lucretia* (1946), *Midsummer Night's Dream* (1960), *Curlew River* (1964), and *The Burning Fiery Furna* (1966) by Benjamin Britten (1913–1976) present different characters with psychological problems [31]. Hans Werner Henze emphasized the conflicts posed by the progress of society in *Boulevard Solitude* (1952), *Elegy for Young Lovers* (1961), and *Die Bassariden* (1966). He also presented psychological reactions resulting in mental disorders in *Wir erreichen den Fluss* (We Come to the River, 1976), where a patient in a psychiatric clinic suffers from progressive functional blindness (a similar disorder had already been depicted in *Il Furioso sull'isola di San Domingo* by Gaetano Donizetti). Further examples of psychological blindness were represented in *Richard Coeur de Lion* (1784) by André Ernest Modeste Grétry (1741–1813), *Le Deux Aveugles* (*The two blind men*, 1855) by Jacques Offenbach (1819–1880), and *Opéra d'Aran* (1962) by Gilbert Bécaud (1927–2001) [32].

Depression was represented in *Jakob Lenz* (1977–1978) and the conflict between sexual drives and hysteria was the subject of *Das Gehege* (The Cage, 2006) by the German composer Wolfgang Rhim. Davies Peter Maxwell (1934–2016) in *Eight songs for a Mad King* (1969), the Danish Bo Holten in *Livlægens Besøg* (*The visit of the Royal Physician*, 2008), and the German Detlev Glanert in *Nijinskys Tagebuch* (2008) addressed the issue of schizophrenia.

Some more recent operas have dealt with neuroscientific themes, such as *The Man Who Mistook His Wife for a Hat* (1986) by the English composer Michael Nyman, who adapted a famous book by the neurologist Oliver Sacks (1933–2015) that deals with a case of prosopagnosia. Nyman himself, in *Facing Goya* (2000) refers to craniology or craniometry as a way to demonstrate whether the characteristics of a genius (as in the case of Goya) are linked to the size of the skull. This theme is also presented in an opera by the Italian Fabrizio de Rossi Re entitled, *Cesare Lombroso,* or the body as a moral principle (2001), which is inspired by the founder of criminal anthropology, author of debatable theories much discussed within and outside the neuroscientific field.

Cognitive impairment or dementia has been represented in several operas, including *King Lear* by William Shakespeare, which has been adapted into opera by several composers: *The Lear or Private Life* (1820) by William Thomas Moncrieff (1794–1857), *King Lear* (2009) by Antonio Cagnoni (1828–1896), and *Lear* (1978) by Aribert Reimann. More recent operas dealing with dementia are *The Lion's Face* (2010) by Elena Langer and Glyn Maxwell, and *Love Hurts* (2016) by Nicola Moro, an opera on the hospitalization in an asylum for "libertine dementia" of the *Marquis De Sade* (Donatien-Alphonse-François de Sade (1740–1814) [4, 33–35] (Table 4.1).

The ethical issues associated with consciousness impairment following serious neurological disorders such as coma or locked-in syndrome due to head trauma or stroke have been explored by the opera director Romeo Castellucci in his 2014 production of Berlioz' *Orphée et Eurydice*, where the drama was set in a neurointensive care unit [36]. This production was a fascinating example of how opera can

Table 4.1 Different forms of representation of madness in opera from the 17th to the 19th century

Mythological madness		
Opera	Year	Composer
Il ritorno di Ulisse in Patria	1640	Claudio Monteverdi
Dido	1641	Francesco Cavalli
L'Egisto	1643	Francesco Cavalli
Atys	1676	Jean-Baptiste Lully
Totila	1677	Giovanni Legrenzi
Roland	1685	Jean-Baptiste Lully
Orlando generoso	1691	Agostino Steffani
Mèdée	1693	Marc-Antoine Charpentier
Orlando furioso	1713	Antonio Vivaldi
Orlando finto pazzo	1714	Antonio Vivaldi
Orlando	1733	George Frideric Haendel
Roland	1778	Niccolò Piccinni
Iphigénie en Tauride Orlando paladin	1779	Christoph Willibald Gluck
	1782	Franz Joseph Haydn
Simulation madness		
Opera	Year	Composer
La finta pazza Licori	1627	Claudio Monteverdi
La finta pazza	1641	Francesco Sacrati
Arcifanfano - Re dei matti	1749	Baldassare Galuppi
Italiana in Algeri	1813	Gioacchino Rossini
I pazzi per progetto	1830	Gaetano Donizetti
Commedia dell'arte Madness		
The Ephesian Matron	1769	Charles Dibdin
La finta giardiniera	1786	Wolfgang Amadeus Mozart
Nina, o sia La pazza per amore	1786	Giovanni Paisiello
Remorse of conscience madness		
Semiramide	1823	Gioacchino Rossini
L'esule di Roma	1828	Gaetano Donizetti
Nabucco	1842	Giuseppe Verdi
I Masnadieri (1847)	1847	Giuseppe Verdi
Nabucco	1853	Giuseppe Verdi
La Traviata	1867	Giuseppe Verdi
Boris Godunov	1869–1872	Modest Petrovič Musorgsky
External agents madness		
Il Pirata	1827	Vincenzo Bellini
La Muette de Portici (1828) by	1828	Daniel Auber
Amorous madness		
Anna Bolena	1830	Gaetano Donizetti
Gemma di Vergy	1834	Gaetano Donizetti
Lucia di Lammermoor	1835	Gaetano Donizetti
Linda di Chamounix	1842	Gaetano Donizetti
Madness as conversional disorder		
Torquato Tasso	1833	Gaetano Donizetti
Il furioso all'isola di San Domingo	1833	Gaetano Donizetti
Maria Padilla	1841	Gaetano Donizetti

effectively convey emotional states and increase public awareness for neurological issues [37, 38].

4.4 Conclusions

From its beginning, opera has frequently represented mental illness, madness, and other neurological conditions. Evaluating how these disorders have been depicted in operas over the centuries allows us to shed a fascinating light on the cultural, social, and medical attitudes and beliefs towards psychiatric and neurological disorders in various historical periods.

References

1. Horden P. Music as medicine. The history of music therapy since antiquity. London: Routledge, Taylor & Francis Group; 2000.
2. Estes JW. The changing role of the physicians in opera. Opera Q. 1993;10(2):143–55.
3. Hutcheon L, Hutcheon M. Opera. The art of dying. Cambridge: Harvard University Press; 2004.
4. Lorusso L, Franchini AF, Porro A. Opera and neuroscience. Prog Brain Res. 2015;216:389–409.
5. Vannoni G. A un dottor della mia sorte. Bologna: Pendragon; 2017.
6. Soriano JB. On doctors and their operas. A critical (and lyrical) analysis of medicine in opera. Chest. 2018;154(2):409–15.
7. Cordingly J. Disorders heroes in opera. A psychiatric report. London: Plumbago Books; 2015.
8. Matthews BR. Bravo! Neurology at the opera. In: Bogousslavsky J, Hennerici MG, Bäzner H, Bassetti C, editors. Neurological disorders in famous artists – part 3. Basel: Karger; 2010. p. 119–29.
9. Bianconi L. Il Seicento. Torino: EDT; 1991.
10. Masiello M. Dalla dottrina dell'êthos alla teoria degli affetti. In: Conservatorio di Musica "Niccolò Piccinni". Accordatura e temperamento: cronaca di un 'compromesso storico'. Bari: Conservatorio di Musica "Niccolò Piccinni" Bari; 2016. p. 25–30.
11. Grier F. Thoughts on Rigoletto. Int J Psychoanal. 2011;92:1541–59.
12. Vecchi O. L'Amfiparmaso. Comedia Harmonica. Venetia: Gardano; 1597.
13. Garzoni T. La piazza vniversale di tvtte le professioni del mondo, e nobili et ignobili. In Venetia: Somascho; 1585.
14. Garzoni T. L'hospidale de' pazzi incurabili. Venetia: Somascho; 1586.
15. Esse M. Rossini's noisy bodies. Camb Opera J. 2009;21(1):27–64.
16. Loges BF. Heiliger Wahnsinn auf der Bühne: die Figur der Hysterika in der Belcanto-Oper. München: epodium; 2010.
17. Pugliese RM. The origins of Lucia di Lammermoor's cadenza. Camb Opera J. 2004;16(1):23–42.
18. Hutcheon L, Hutcheon M. Death drive: Eros and Thanatos in Wagner's Tristan und Isolde. Camb Opera J. 1999;11(3):267–93.
19. Brunner J, Hirsch T, Steger F. Inszenierung von akzentuierten Persönlichkeitszügen in der Barockoper. Opera buffa Arcifanfano – Re dei matti (1749). Nervenarzt. 2016;87:528–33.
20. Matsumoto N. The operatic mad scene: its origins and early development up to C. 1700. Thesis Goldsmiths' College University of London; 2004.
21. Pugliese RM. Il topos della follia nell'opera. Prospettive critiche. In: Sala E, editor. Mad scenes & Songs. Quaderno delle notti Malatestiane, vol. 115; 2002. p. 229.
22. Sala E, Ashbrook W. Women crazed by love. An aspect of romantic opera. Opera Q. 1994;10:19–41.

23. Erfurth A, Hoff P. Mad scenes in early 19th-century opera. Acta Psychiatr Scand. 2000;102:310–3.
24. Cambioli L, Bellelli G, Clerici M, Cesana G, Riva MA. "Nabucco" by Giuseppe Verdi. A case of *Delirium* in an Italian Romantic Opera. Eur Neurol. 2017;77:180–5.
25. Gallini C. La sonnambula meravigliosa. Magnetismo e ipnotismo nell'Ottocento Italiano. Milano: Feltrinelli; 1983. p. 43.
26. Finger S, Sironi VA, Riva MA. Somnambulism in Verdi's Macbeth and Bellini's La Sonnambula: opera, sleepwalking, and medicine. Prog Brain Res. 2015;216:357–88.
27. Riva MA, Mazzocchi C, Cesana G, Stanley F. Il sonnambulo' by Michele Carafa: a forgotten romantic opera with sleepingwalking. Eur Neurol. 2016;76(5–6):210–1.
28. Brigo F, Lattanzi S. Sleep disorders in Cimarosa's "the secret marriage". Neurol Sci. 2028;39(10):1805–6.
29. Göbel CH, Göbel A, Göbel H. "Compulsive plague! Pain without end". How Richard Wagner played out his migraine in the opera Siegfried. BMJ. 2013;347:f6952.
30. Göbel A, Göbel CH, Göbel H. Phenotype of migraine headache and migraine aura of Richard Wagner. Cephalalgia. 2014;34:1004–11.
31. Scull A. Some reflections on madness and culture in the post-war world. Hist Psychiatry. 2014;25:395–403.
32. Aydin P, Ritch R, O'Dwyer J. Blindness and visual impairment in opera. EJO. 2018;28:6–12.
33. Fuller D. Dementia at the opera. The Lion's Face Opera Qaurterly. 2011;27:509 21.
34. Grandi R. Adattamenti, riscritture, burlesques (1681–1860). King Lear dopo Shakespeare. Roma: Aracne; 2013.
35. Raz C. Music, theater and the moral treatment: the *Casa dei Matti* in Aversa and Palermo. Lab Ital. 2017;20. journals.openedition.org/laboratoireitalien/1581
36. Brigo F. Orpheus' descent into the neurointensive care unit. JAMA. 2018;319(4):328.
37. Trimble M, Letellier RI, Hesdorffer D. Sudden death in opera: love, mortality and transcendence on the lyric stage. Newcastle upon Tyne: Cambridge Scholars Publishing; 2021.
38. Pridmore SA, Auchincloss S, Soh NL, Walter GJ. Four centuries of suicide in opera. Med J Aust. 2013;199:783–6. https://doi.org/10.5694/mja13.10724.

Neuro-Musicology of Opera

5

Francesco Brigo, Alessandro Porro, and Lorenzo Lorusso

5.1 Introduction

Music is a cultural aspect present in all human societies, where it plays an important part in people's way of life. It is the art of combining and organizing sounds in time for beauty of form or emotional expression. There is a huge variety in the types and styles of music, which can be differentiated by the instrumental and vocal techniques adopted and by its constitutive elements: melody, harmony, rhythm, and timbre.

With its characteristic and fascinating combination of music and language to be sung, opera represents a unique artistic genre that, in its theatrical performance, incorporates a variety of different arts, such as acting, scenery, costume, and sometimes dance. The variety of opera's components has exerted a great fascination on spectators all over the world since its creation in the seventeenth century, as the most typical product of baroque Italian creativity, which did not fear contamination with different artistic genres. Because of this intrinsic complexity in its structure, the effects of opera on human physiology and the functioning of the nervous system should be evaluated on their own.

However, compared to the wealth of research on the effects of music on the brain [1–3], the number of studies devoted to opera is surprisingly scarce. The reasons for

F. Brigo
Department of Neurology, Hospital of Merano-Meran (SABES-ASDAA), Merano-Meran, Italy

A. Porro
History of Medicine, University of Milan, Milan, Italy

L. Lorusso (✉)
Neurology Unit, ASST-Lecco, Merate Hospital, Merate, Italy
e-mail: l.lorusso@asst-lecco.it

L. Lorusso et al. (eds.), *Effects of Opera Music from Brain to Body*, Neurocultural Health and Wellbeing,
https://doi.org/10.1007/978-3-031-34769-6_5

such a discrepancy have never been investigated. Undoubtedly, disentangling the contribution of each component of opera to the functions of the nervous system is more challenging than focusing on unsung music devoid of any theatrical element.

In this chapter, we provide an overview of the main studies available in the scientific literature that have investigated the effects of opera on the human brain and its possible application for the treatment of different health problems.

5.2 Opera Performance: Self-Identification on the Stage and in the Brain

Attending an opera performance in a theatre or simply listening to it while sitting at home can be a hugely emotional experience. Such an emotional response can be so intense as to engender a real passion among opera fans [4], which in turn contributes to the perpetuation of opera performances, despite their enormous costs and organizational challenges [5]. This passion can lead to very heated discussions between fans of the genre, not unlike those that occur among supporters of rival football teams.

A famous literary example of such a profoundly emotional response to opera and the related physicality of this experience is represented by Gustave Flaubert's *Madame Bovary* (1857). In the book, the protagonist attends a performance of Gaetano Donizetti's *Lucia di Lammermoor* and identifies herself with the heroine's love passion, whereas her dull husband irritates her with ignorant questions, unable to follow the plot. While listening to the music, "She let herself be lulled by the melodies, feeling her whole being vibrate, as if the bows of the violins were playing on her own taut nerves. Her avid gaze took in every detail of the costumes, the set, the characters, the painted trees that quivered when anyone walked past, the velvet caps, the cloaks, the swords, all that imagined reality which the music, like the atmosphere of some other world, quickened into life" [6] (Fig. 5.1).

Emma, the prototypical example of a nineteenth-century hysteric woman, "recognized that same rapture, that same anguish that had brought her so close to death. The voice of the soprano seemed simply an echo of Emma's own heart, and the illusion that held her spellbound a part of Emma's very own life" [6]. As recently proposed within the framework of "theory of mind" (i.e., the simulative capacity to make inferences about the mental states of others [7]), attending a musical or opera performance would represent a sort of "affective sandbox," allowing spectators to pursue emotion exploration and hypothesis testing in safe environments, just like in the case of Emma Bovary. Thus, echoing Gustave Flaubert (1821–1880), every opera fan could exclaim *"Madame Bovary c'est moi!"* [8]: now as then, the dramatic character of opera amplifies its intrinsic emotional resonances, allowing the spectator to identify with the character on the stage. Such identification is not only metaphorical but, to some extent, actually occurs in the brains of the spectators during the audiovisual perception of an opera performance.

One study has investigated whether an opera performance can induce mirror neuron activity in the spectators [9].

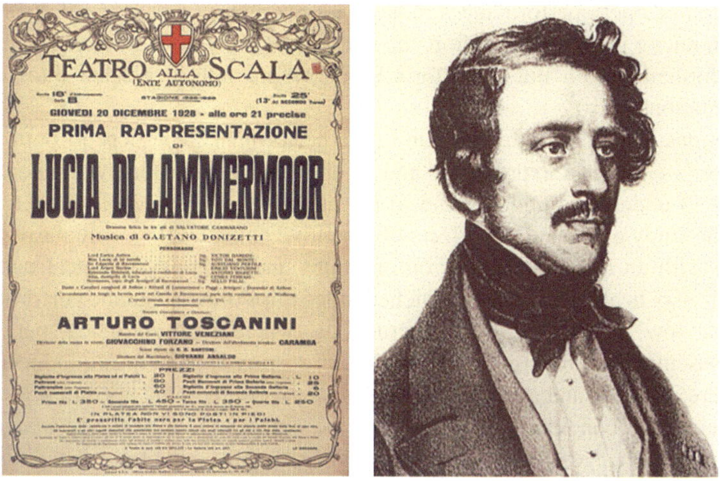

Fig. 5.1 Poster of "Lucia di Lammermoor" for "Teatro alla Scala", December 29, 1928, conductor Arturo Toscanini (1868–1957). Gaetano Donizetti (1797–1848), etching (1830 circa)

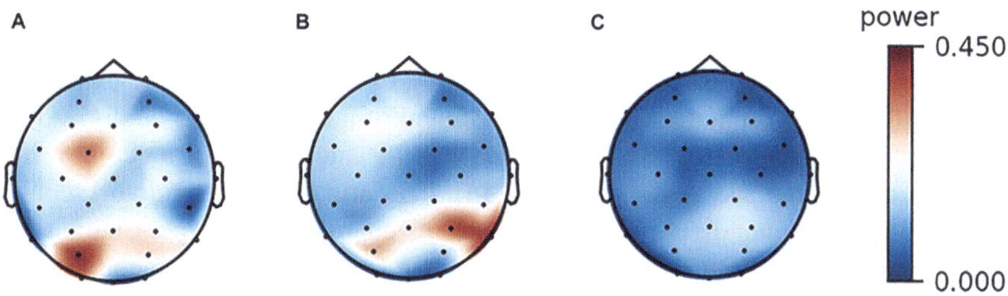

Fig. 5.2 Among singers who performed in opera and concerts, the alpha-band powers in the audiovisual (**c**) and auditory conditions (**b**) are significantly lower than those in the resting condition (**a**). Tanaka S. Mirror Neuron Activity During Audiovisual Appreciation of Opera Performance. Front Psychol 2021; 11: 563031 DOI: 10.3389/fpsyg.2020.563031

The human mirror system consists of neurons (so-called mirror neurons) that activate during self-performances and while observing others performing similar actions [10]. This neuronal network, which represents the neurophysiological mechanism underlying the coupling of perception and action, involves the ventral premotor cortex and inferior parietal lobule [11]. The study showed that in singers listening to and watching an opera performance, the alpha frequency band power of the electroencephalogram changed compared to the resting condition by selectively decreasing in the fronto-centro-parietal areas [9] (Fig. 5.2).

Conversely, while only listening to an operatic aria, the alpha-band power did not change. Although this study did not include functional magnetic resonance imaging, source localization, or network analysis, these findings suggest that audiovisual perception of opera performances induces mirror neuron activity in the audience. Through mirror neuronal activation, the opera singer on the stage shares an embodied performance with the spectator's brain, where it is translated into internal representations.

Emotions evoked by attending an opera performance can depend both on the performer's emotional expressivity and on the observer's empathy, defined as the capacity to understand and respond to the affective experience of another person [12]. One study evaluated the effects of cognitive empathy (i.e., voluntarily empathizing) with an opera performer on emotions and their physiological correlates [13]. Fifty-six healthy volunteers watched video clips of two operatic performances, following low- or high-empathy instructions, whereas physiological parameters (cardiovascular, electrodermal, and respiratory responses) and music-induced emotions were recorded and analyzed. Audiovisual exposure to the operatic excerpts in a high-empathy condition facilitated specific emotions and was associated with physiological changes. Compared to the low-empathy condition, listening to an operatic aria with sad content facilitated the emotion of nostalgia and skin conductance level, whereas exposure to the aria with happy content led to an emotion of power and increased respiration frequency. These findings suggest that cognitive empathy influences the psychophysiology of emotion during music listening in general and attendance at operatic performances in particular.

Of note, the mechanisms underlying the activation of the human mirror neuron system induced by emotions evoked by listening to music appear to be experiential rather than representational. The involvement of mirror neurons could lead to motor representations of emotions by recruiting the anterior insula, a neural relay between the limbic and motor systems [14]. The mirror neuron network could therefore play a pivotal role in the neuronal mechanisms underlying language, motor functions, and music and in their intrinsic ability to express meaning and emotions. As recent behavioral, electrophysiological, and neuroimaging studies have revealed, the neuroanatomic correlates of the functional and structural relationships between music and language are extremely complex. The linguistic and musical dimensions of songs (and, by extension, of operatic arias) undergo processing by overlapping neuronal networks in frontal and temporal regions, which are mutually influenced by elaborate interactions [5]. However, although probably mediated by overlapping neuronal circuits, the processing of semantic aspects of language and harmonic aspects of music occurs at varying degrees of integration (and separation) when listening to opera [15], with the language being processed before the music [16, 17]. This neurophysiological finding appears to put an end to the long quarrel about the presumed superiority of the music over the text (musicocentrism) or vice versa (logocentrism) [18]. Language and music represent complex entities consisting of specific elements, each probably involving different types of neuronal processing. However, in the spectator's perception, lyrics and tunes appear to be linked in a natural and yet mysterious way. As the composer and conductor Pierre Boulez (1925–2016) wrote, "the text is the center and the absence of the musical piece" [19].

5.3 "Fremo, smanio, avvampo, tremo" (I Tremble, Yearn, Flush, Tremble): Opera and Its Psychophysiological Effects Opera and Its Psychophysiological Correlates Opera and the Autonomic Nervous System

Each opera component (music, plot, and acting performance) can have a specific impact on the emotions it can evoke and on their psychophysiological correlates in the spectator. The role of any of these multimodal sensory stimuli has been evaluated in a study exposing 37 healthy volunteers to three different experimental conditions: (1) listening to excerpts from Giacomo Puccini's *Tosca* (1904), (2) music relistening after learning the plot, and (3) music relistening while watching a subtitled film of the acting performance [20]. Several physiological parameters (including cardiovascular, electrodermal, and respiratory responses) and emotion descriptors were recorded and analyzed. Music listening alone was associated with increased heart frequency, but slower respiration frequency and reduced skin conductance. Listening again to the same music after having learned the tragic plot was associated with reduced positive emotions (peacefulness, joyful activation) and increased sadness, decreased respiratory sinus arrhythmia, and increased power in the low-frequency band of heart rate variability. Watching the acting performance increased emotional arousal, with emotions reported to change from "less positive"/"sad" to "transcendent" and continued high autonomic arousal. Despite some limitations, the main being the laboratory rather than a naturalistic setting, this study found that different approaches to opera (music listening, learning the plot, and watching the acting performance) have specific effects on music-induced emotions and their physiological correlates. Thanks to these subconscious effects on the autonomic nervous system, opera could be used to improve psychophysical performances. As an example, engaging children with opera has been increasingly proposed as an educational activity to improve their perception and analytical skills [21–24] (Fig. 5.3).

5.4 "Opera Therapy": Exploiting the Treating Potential of Opera

The effects of opera on emotions and their physiological correlates also represent the rationale for using it as a way to treat health problems. Listening to music can be a pleasant experience with positive effects on quality of life, both in healthy subjects and in people with health problems. Music medicine consists of passive exposure to pre-recorded music administered by health personnel [25], whereas music therapy is an active intervention using "music interventions to accomplish individualized goals within a therapeutic relationship by a credentialed professional who has completed an approved music therapy program" [26]. In the field of neurology, there is some evidence of a beneficial effect of music on many neurological conditions, including movement disorders, senile dementia, epilepsy, and sleep disorders

Fig. 5.3 The title page, by German artist Adolf Hohenstein (1854–1928), of the first edition of the piano score to Puccini's opera Tosca, published between Spring 1896 and Autumn 1899) by G. Ricordi. Libretto of Tosca, published by G. Ricordi (1899), cover illustrated by Alfredo Montalti (1858–1928)

[27–32]. However, studies evaluating the role of "opera therapy" for the treatment of health problems are scarce.

In modern times, the positive effects of music therapy for neuropsychiatric diseases were first exploited during the great depression by the US government, under the presidency of Franklin D. Roosevelt (1882–1945) and involving the musician Nikolai Sokoloff (1866–1965). The Federal Music Project consisted of several musical events aimed at patients with different neuropsychiatric disorders, under the supervision of professionals in the medical and musical fields [33]. The project used different musical genres, including "Grand Opera, Operetta, Opera Comics and Chamber Opera," to facilitate positive emotions among participants [34] (Fig. 5.4).

The use of opera pieces for the treatment of dementia has been proposed as an effective way of evoking positive autobiographical memories through music in patients who have been familiar with this musical genre in the past. According to this hypothesis, opera could therefore act as a sort of Proustian *madeleine,* able to unlock recollection by evoking emotions and fragments of memories from the participants' past lives. One study has evaluated the preservation of musical memory in healthy elderly people and patients affected by Alzheimer's disease [35]. In dementia patients, only one operatic piece (the melodies of the bridal chorus *"Treulich geführt"* from Richard Wagner's *Lohengrin* [1850]) was used out of 12 musical excerpts. The reason for the choice of this piece was not reported by the authors; probably it was selected as it is a march played for the bride's entrance at many

Fig. 5.4 Weitzman, M. and Federal Music Project, U. S. *W.P.A. Federal Music Project information given regarding locations of W.P.A. schools and all subjects taught*. New York, None. [New York: federal art project, between 1936 and 1941] [Photograph] Retrieved from the Library of Congress, Verdi, G. and Federal Music Project, U. S. (1936) *WPA in Ohio Federal Music Project presents Verdi's grand opera "Il Trovatore" Cast of 75: Rudolph Schueller director*. Ohio, 1936. [Ohio: Federal Art Project, or 1937] [Photograph] Retrieved from the Library of Congress

formal weddings throughout the Western world. Conversely, in the study conducted in cognitively intact subjects, opera music by Gioachino Rossini and Georges Bizet appeared as excerpts from the Ouvertures from *Il Barbiere di Siviglia* (*The Barber of Seville*, 1816) and *La Gazza Ladra* (*Thieving Magpie*, 1817) and as the suite from *Carmen*. The choice of instrumental operatic melodies rather than arias was aimed at avoiding interference from the text.

Music therapy has been widely advocated as an effective way of engaging patients with dementia and their caregivers in purposeful and enjoyable activities, reducing behavioral and psychological symptoms [36–41], and providing them with the opportunity to become re-engaged within a community [42]. All these aspects undoubtedly apply to listening to opera or attending an opera performance. However, active involvement in the creation of an opera performance could also prove extremely beneficial for people with dementia.

In a creative musical project involving professional singers and other staff from the Scottish Opera, patients with dementia and their caregivers were actively involved in the development, writing, design, and performance of a short opera based on their own experiences of love [43]. The project, funded through a charitable grant, ran for less than one year, with weekly practice sessions involving a wide range of activities reflecting the different components of opera (singing, acting, dancing, composing, art projecting, and set designing), which were followed by a series of performances. The collaborative nature of this project was greatly appreciated by all participants and had positive outcomes, such as improved confidence, an increased sense of belonging to a group, improved physical strength and social skills, as well as improved satisfaction and self-affirmation.

A controlled study evaluated the effect of traditional Chinese opera on the cognitive functions in elderly patients with mild and moderate dementia [44]. In the active group, 21 Chinese patients with dementia and an interest in traditional opera were invited to learn to sing, memorize the lyrics, tunes, and movements, and eventually perform a traditional Chinese opera. The intervention lasted 12 weeks. Compared to controls, participants in the project had statistically significant improvements in their cognitive functions, as measured by the Mini-Mental State Examination, the Chinese version of the neuropsychiatric inventory, and the Quality of Life in Alzheimer's disease. These findings were explained by the facilitation of long-term memory recall and expression of emotions through active engagement with music rooted in the participants' cultural background and life experiences.

5.5 Conclusions

Opera includes several components that make it an all-encompassing artistic form and a source of multimodal sensory stimulation for the human brain. Thanks to the profound interaction between its various components, opera can evoke emotions and heighten their expression as signified by the text and presented in the staged action. The exact mechanisms underlying the intense emotional response it evokes in the spectator are still far from being fully elucidated. They probably include the activation of the human mirror neuron system and the cerebral areas underlying language, action, and music and their ability to express meaning and emotions. The effects of opera on music-induced emotions and their physiological correlates could be used for the treatment of health problems. Despite some intriguing data, the studies evaluating the effectiveness of "opera therapy" are scarce and almost exclusively confined to the treatment of dementia. Future research in this field should clarify the psychophysiological effects of opera as a whole artistic experience and each of its components. Further studies are warranted to identify which health disorders may especially benefit from "opera therapy" and whether there are differences in clinical effectiveness compared to other music therapies.

References

1. Edwards J. The use of music in healthcare contexts: a select review of writings from the 1890s to the 1840s. Voices. 2008;8(2) https://doi.org/10.15845/voices.v8i2.
2. Thaut MH, Hoemberg V, editors. Handbook of neurologic music therapy. Oxford: Oxford University Press; 2016.
3. Baker FA. Therapeutic songwriting. Developments in theory, methods and practice. Basingstoke: Palgrave Macmillan; 2015.

4. Matthews BR. Bravo! Neurology at the opera. Front Neurol Neurosci. 2010;27:119–29. https://doi.org/10.1159/000311196.
5. Schon D, Gordon RL, Besson M. Musical and linguistic processing in song perception. Ann N Y Acad Sci. 2005;1060:71–81.
6. Flaubert G. Madame Bovary. Provincial manners. Oxford: Oxford University Press; 2004.
7. Vandervert L. The evolution of theory of mind (ToM) within the evolution of cerebellar sequence detection in stone tool making and language: implications for studies of higher-level cognitive functions in degenerative cerebellar atrophy. Cerebellum Ataxias. 2019;6:1. https://doi.org/10.1186/s40673-019-0101-x.
8. Descharmes R. Flaubert. Sa vie, son caractère et ses idées avant 1857. Montréal: Lacoste; 2014.
9. Tanaka S. Mirror neuron activity during audiovisual appreciation of opera performance. Front Psychol. 2021;11:563031. https://doi.org/10.3389/fpsyg.2020.563031.
10. Rizzolatti G, Fogassi L. The mirror mechanism: recent findings and perspectives. Philos Trans R Soc B Biol Sci. 2014;369:369. https://doi.org/10.1098/rstb.2013.0420.
11. Oztop E, Kawato M, Arbib MA. Mirror neurons: functions, mechanisms and models. Neurosci Lett. 2013;540:443–55.
12. Decety J, Jackson PL. A social neuroscience perspective on empathy. Curr Dir Psychol Sci. 2006;15:54–8.
13. Miu AC, Baltes FR. Empathy manipulation impacts music induced emotions: a psychophysiological study on opera. PLoS One. 2012;7(1):e30618. https://doi.org/10.1371/journal.pone.0030618.
14. Molnar-Szakacs I, Overy K. Music and mirror neurons: from motion to "e"motion. Soc Cogn Affect Neurosci. 2006;1(3):235–41.
15. Sammler D, Baird A, Valabrègue R, Clément S, Dupont S, Belin P, Samson S. The relationship of lyrics and tunes in the processing of unfamiliar songs: a functional magnetic resonance adaptation study. J Neurosci. 2010;30(10):3572–8.
16. Besson M, Faïta F, Peretz I, Bonnel AM, Requin J. Singing in the brain: independence of lyrics and tunes. Psychol Sci. 1998;9(6):494–8.
17. Besson M, Schön D. Comparison between language and music page. In: Peretz I, Zatorre RJ, editors. The cognitive neuroscience of music. New York: Oxford University Press; 2003. https://doi.org/10.1093/acprof:oso/9780198525202.003.0018.
18. Gorlée DL, editor. Song and significance. Virtues and vices of vocal translation. Amsterdam: Rodopi; 2005.
19. Boulez P. Relevés d'apprenti. Paris: Éditions du Seuil; 1966.
20. Baltes FR, Avram J, Miclea M, Miu AC. Emotions induced by operatic music: psychophysiological effects of music, plot, and acting: a scientist's tribute to Maria Callas. Brain Cogn. 2011;76(1):146–57. https://doi.org/10.1016/j.bandc.2011.01.012.
21. Burrack F, Maltas C. Engaging elementary age children with opera. Appl Res Mus Educ. 2002;25:82–9.
22. Harris M. Music and the young mind. Enhancing brain development and engaging learning. Lanham: Rowman & Littlefield Education; 2009.
23. Dickinson P. Music education in crisis. The Bernarr rainbow lectures and other assessments. Suffolk: Boydell & Brewer; 2013.
24. Taylor CL. Opera education in the upper elementary music classroom. Oxford: University of Mississippi; 2015.
25. Stegemann T, Geretsegger M, Phan Quoc E, Riedl H, Smetana M. Music therapy and other music based interventions in pediatric health care: an overview. Medicines. 2019;16(81):25. https://doi.org/10.3390/medicines6010025.
26. AMTA American music therapy Association Official Definition of Music Therapy. 2005. https://www.musictherapy.org/about/musictherapy/. Accessed 18.12.2021.
27. Devlin K, Alshaikh JT, Pantelyat A. Music therapy and music based interventions for movement disorders. Curr Neurol Neurosci Rep. 2019;19(11):83.

28. Irish M, Cunningham CJ, Walsh JB, Coackley D, Lawlor BA, Robertson IH, Coen RF. Investigating the enhancing effect of music on autobiographical memory in mild Alzheimer's disease. Dement Geriatr Cogn Disord. 2006;22(1):108–20.
29. Petrovsky DV, Ramesh P, McPhillips MV, Hodgson NA. Effects of music interventions on sleep in older adults: a systematic review. Geriatr Nurs. 2021;42(4):869–79.
30. Sousa L, Neves MJ, Moura B, Schneider J, Fernandes L. Music-based interventions for people living with dementia, targeting behavioral and psychological symptoms: a scoping review. Int J Geriatr Psychiatry. 2021;36(11):1664–90.
31. Lin LC, Lee WT, Wang CH, Chen HL, Wu HC, Tsai CL, Wei RC, Mok HK, Weng CF, Lee MW, Yang RC. Mozart K 448 acts as a potential add-on therapy in children with refractory epilepsy. Epilepsy Behav. 2011;20(3):490–3.
32. Bodner M, Turner RP, Schwacke J, Bowers C, Norment C. Reduction of seizure occurrence from exposure to auditory stimulation in individuals with neurological handicaps: a randomized controlled trial. PLoS One. 2012;7(10):e45303. https://doi.org/10.1371/journal.pone.0045303.
33. Davis WB. Music therapy practice in New York City: a report from a panel of experts, March 17, 1937. J Mus Ther. 1997;34:68–81.
34. Soibelman D. Therapeutic and industrial uses of music. New York: Columbia University Press; 1948.
35. Cuddy LL, Sikka R, Vanstone A. Preservation of musical memory and engagement in healthy aging and Alzheimer's disease. Ann N Y Acad Sci. 2015;1337(1):223–31.
36. Raglio A, Bellelli G, Mazzola P, Bellandi P, Giovagnoli AR, Farina E, Stramba-Badiale M, Gentile S, Gianelli MV, Ubezio MC, Zanetti O, Trabucchi M. Music, music therapy and dementia: a review of the literature and recommendations of the Italian psychogeriatric association. Maturitas. 2012;72:305–10.
37. Raglio A, Bellelli G, Traficante D, Gianotti M, Ubezio MC, Gentile S, Villani D, Trabucchi M. Efficacy of music therapy treatment based on cycles of sessions: a randomized controlled trial. Aging Ment Health. 2010;14(8):900–4.
38. Camic PM, Meeten F, Williams M. Does a "singing together group" improve the quality of life of people with a dementia and their carers? A pilot study evaluation. Dementia. 2013;12(2):157–76.
39. Heeyoun KC. The effects of group singing on quality of life and affect of people with dementia: a randomized controlled trial. Philadelphia: Temple University; 2016.
40. Fukui H, Arai A, Toyoshima K. Efficacy of music therapy in treatment for the patients with Alzheimer's disease. Int J Alzheimer Dis. 2012;4:1–6. https://doi.org/10.1155/2012/531646.
41. Suzuki M, Kanamori M, Watanabe M. Behavioral and endocrinological evaluation of music therapy for elderly patients with dementia. Nurs Health Sci. 2004;6:11–8.
42. Wood S. Chalfont lodge choir: hearth of a home and community. J Dementia Care. 2007:22–5.
43. Mc Cabe L, Greasley-Adams C, Goodson K. "What I want to do is get half a dozen of them and go and see Simon Cowell": reflecting on participation and outcomes for people with dementia taking part in a creative musical project. Dementia. 2015;14(6):734–50.
44. Chen X, Li D, Xu H. Effect of traditional opera on older adults with dementia. Geriatr Nurs. 2020;41(2):118–23.

The Singing Brain: Words and Music in the Opera

6

Francesco Brigo, Alessandro Porro, and Lorenzo Lorusso

6.1 Introduction

Opera singers know well how to produce sounds with their voice; modulate their pitch, timbre, and volume; and accompany them with the most appropriate gestures to convey dramatic meaning and emotional intensity to the audience. What most opera singers probably do not know is that each of these actions is precisely orchestrated and carefully organized by a unique musical conductor and an extraordinary director. Unlike those guiding singing and acting on stage, this conductor and director always remain behind the curtains, completely invisible to anyone, hidden within the bodies of singers. The organ that incessantly and meticulously plans and supervises the motor activities of singing and acting is the human brain. Neuroanatomic substrates of opera singing include cortical motor areas controlling the functions of the larynx and orofacial articulators (including the lip, jaw, and tongue), the auditory cortex, and subcortical and cerebellar networks involved in implicit motor memory [1]. What is less known is that the complex interaction of music, text, and performance that characterizes opera can in turn influence the functioning of the brain. In this chapter, we focus on the fascinating interplay between opera and the nervous system, trying to elucidate the effects of opera singing on neurophysiology.

F. Brigo
Department of Neurology, Hospital of Merano-Meran (SABES-ASDAA), Merano-Meran, Italy

A. Porro
History of Medicine, University of Milan, Milan, Italy

L. Lorusso (✉)
Neurology Unit, ASST-Lecco, Merate Hospital, Merate, Italy
e-mail: l.lorusso@asst-lecco.it

L. Lorusso et al. (eds.), *Effects of Opera Music from Brain to Body*, Neurocultural Health and Wellbeing,
https://doi.org/10.1007/978-3-031-34769-6_6

6.2 Historical Notes on the Role of Music and Voice in Early Operas

The earliest studies on vocal voice and sound can be found in some works by Hippocrates (*Aphorisms*, First and Second of the *Epidemics*) [2] and in the Aristotelian *Problemata* [3, 4].

These studies focused on the relation between vocalization and animate motion, with medieval commentators of the *Aphorism*, such as Constantinus Africanus (eleventh century) and Taddeo Alderotti (thirteenth century), who applied this doctrine to the production of sounds. During the Middle Ages, there was scarce knowledge on the role of the throat and larynx in phonation, although Dante Alighieri in his *Divine Comedy* recognized the importance of these organs in respiration and speech [5].

In 1562, the Neapolitan physician and philosopher Camillo Maffei Solofra published the *Discourse on the Voice and on the Method of Learning to Sing "from the throat"* ("singing from the throat," in Italian "Cantar di Garganta," was an ornamented style of singing that was popular at the time) [6]. This important book discussed the psychophysiology of singing based on doctrines by Aristotle and Galen. In 1575, the Spanish Juan Huarte de San Juan (1529–1588) published the book *The Examination of Men's Wits* on the psychology of vocational aptitudes and personality assessment, with a section dedicated to the effect on vocalization of the varying qualities of the intellect and of the imaginative faculty in people with various temperaments [7, 8]. At the end of the sixteenth century, the monographs *Physiologia Vocis* (Physiology of voice) and *De Individualitate Sensibili Faciei Humanae Libri Tres* (On the Individuality of the Human Face as Discernible to the Sense, in Three Books), written by the physician Marcantonio Olmi—who was born in Padua and taught in Bologna—were focused on the relation between personality, voice, and speech type [9]. In 1600, Gaspar Bauhin's *Anatomical History of the Larynx, Organs of Voice* was published in Ferrara. In the same year, in Venice, a book with the same title was published by the anatomist Hieronymus Fabricius ab Acquapendente, who in 1601 wrote the treatise *On Speech and Its Organs*. Johannes Jessenius, a pupil of Fabricius, published a similar account of the organs of voice and speech in his *Anatomies Held a Prague* (1601) [10]. In Ferrara, in 1601, Julius Casserius published an illustrated *Anatomical History of the Organs of Voice and Hearing* [11, 12].

At the turn of the century, while these works from different anatomists were published, the "*Florentine Camerata*" or "*Camerata de' Bardi*," an academy of scholars, poets, singers, and other professional and amateur musicians, began its activities, which culminated in 1597 with the creation and representation of the first opera, *Dafne*, with music by Jacopo Peri. On October 6, 1600, at the Pitti Palace in Florence, *Euridice*, an opera with music composed by Peri, was performed for the wedding of Maria of Medici and Henry IV of France; it is the earliest opera, the music of which still survives (Fig. 6.1).

Interestingly, the "*Florentine Camerata*" had among its members Vincenzo Galilei, Galileo's father. Vincenzo Galilei, in his book *Dialogo della musica antica*

Fig. 6.1 The painting represents the *"Camerata de' Bardi"* or *"Camerata Fiorentina". I musicisti del principe Ferdinando de' Medici,* 1685 circa, Anton Domenico Gabbiani (1652–1726), Florence, Galleria Palatina *(Museo degli strumenti musicali)*

et della moderna, and Francesco Patrici in *Della Poetica* refer to a passage from the Aristotelian *Problemata* to illustrate their attempts to reconstruct the ancient Greek musical drama. Their books were devoted to understanding the physiology and pathology of voice and speech, as further elaborated in the commentary of the Paduan physician Peter of Abano [13–17].

In the late sixteenth century, the influence of song and music on "emotions," including various types of melancholy, was explored in an interdisciplinary context and was the subject of inquiry and research in the different fields of religion, philosophy, and medicine. As an example, some plays by William Shakespeare *(King Lear, Pericles, The Tempest, and the Twelfth Night)* illustrate how music was used to temper and even treat mood disturbances according to the ancient doctrine of Galen [18].

Richard Brocklesby in *Reflections on Ancient and Modern Musick, with the Application to the Cure of Disease* (1749) and Richard Browne in *Medicina Musica* (1729) maintained that the benefits of singing arise from the pressure that the lungs, filled with air, exert on the lower abdomen, thereby favorably stimulating the nerves. Both treaties reflect the eighteenth-century interest in the nervous system as mainly responsible for human health [19, 20].

The results of these historical studies on the influence of voice/singing have been subsequently confirmed by further research, showing that singing proves useful in increasing self-confidence. Singing is a physical activity that is both "energizing and relaxing" and can help relieve stress and tension. Singers commonly believe that singing can improve breathing and lung capacity, promote good posture, and enhance positive moods and emotions, and an overall sense of well-being [21].

6.3 Does Opera Change the Brain? Opera Singing and Neuroplasticity

Professional opera singing requires prolonged training, usually beginning in early adulthood, and extensive practice to achieve outstanding artistic results. This combination of training specificity and intensity starting at an early age results in adaptive anatomical and physiological changes in the brain, a process known as neuroplasticity. As this term implies, the human brain is intrinsically able to modify, change, and adapt its structure and function in response to experience [22].

Studies in professional musicians have shown that sensorimotor training over extended periods of time leads to changes in the cortical representation of sensory and motor systems [23, 24], associated with increased gray matter and cortical thickness in auditory, visual, and motor regions [25–27].

Only a few studies have investigated the changes in cerebral structure and function as a response to the professional training of the voice among opera singers. One reason for the paucity of studies evaluating neuroplasticity in the vocal system is that speech motor control and coordination are already fully developed and automatic in adolescence [28].

However, professional opera singers undergo prolonged and highly specialized vocal training throughout their entire careers to account for the peculiarities and difficulties of this artistic genre. This provides a unique opportunity to better investigate the experience-dependent neuroplastic changes induced by vocal training.

One functional magnetic resonance imaging (fMRI) studz evaluated changes in functional brain activation in highly experienced opera singers, conservatory-level vocal students, and non-singers during the singing of an Italian aria [29] (Fig. 6.2).

Thanks to fMRI, it is possible to noninvasively investigate the functioning of the human brain in vivo, by analyzing the changes in hemodynamic signals (i.e., the metabolic rate of oxygen consumption, cerebral blood flow, and cerebral blood volume) after exposure to a specific event (event-related fMRI) [30] or at rest (resting-state fMRI) [31]. These data provide an indirect measure of neuronal activity [32].

Training in opera singing was associated with increased activation of the bilateral primary somatosensory regions, in the somatotopic areas involved in kinesthetic information processing and proprioceptive feedback of the articulation and larynx. Other areas associated with training in vocal skills were subcortical (basal ganglia and thalamus) and cerebellar networks involved in implicit and procedural motor memory, as well as a frontoparietal network (the inferior parietal lobe and bilateral dorsolateral prefrontal cortex) associated with sensorimotor guidance of motor activity. Functional activation in the right primary sensorimotor cortex was found only in professional opera singers. These findings indicate that training in opera singing is associated with increased somatosensory feedback and kinesthetic perception, which are responsible for transforming highly automatic speech motor sequences into highly specialized motor programs. The activation of the non-speech-dominant primary sensorimotor cortex observed only in participants with years of extensive training and performance in opera singing suggests a deeply intertwined pattern of processing of words and melody.

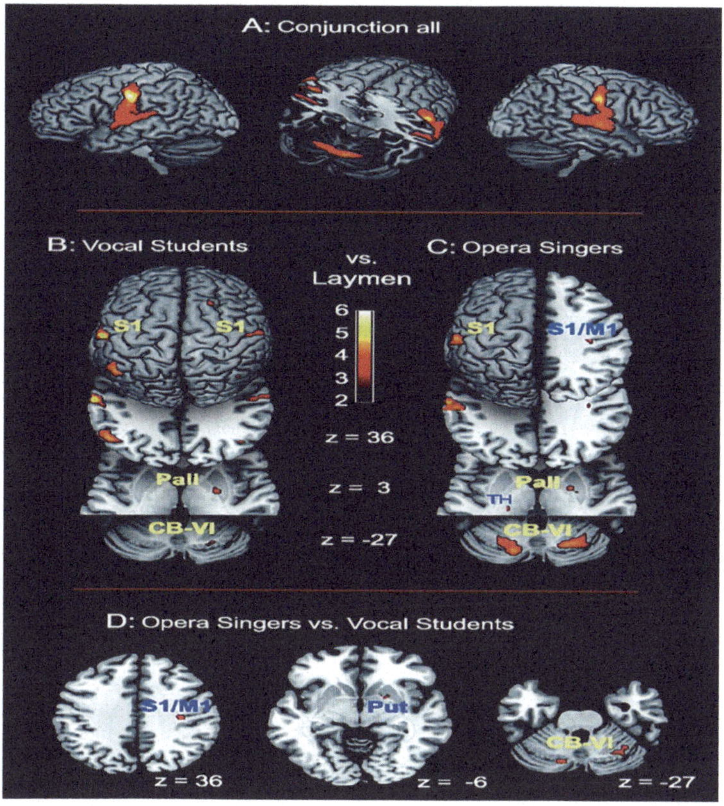

Fig. 6.2 Expertise in classical singing correlates most notably with increased functional activation of bilateral primary somatosensory cortex, representing proprioceptive feedback from the articulators and the larynx, in concert with increased involvement of the cerebellum and implicit motor memory areas at the subcortical level. Kleber B, Veit R, Birbaumer N, Gruzelier J, Lotze M. The brain of opera singers: experience-dependent changes in functional activation. Cereb. Cortex 2010; 20: 1144–1152

In a subsequent study from the same research group, fMRI was used to evaluate the effects of altering somatosensory feedback through laryngeal (vocal-fold) anesthesia in trained singers and non-singers [33].

Compared to non-singers, singers had less impairment in pitch accuracy, and this difference was associated with specific patterns of decreased cortical and subcortical activation in the two groups. The greatest changes were found in the right anterior insula and sensory-motor areas, emphasizing their role in experience-dependent modulation of feedback and feedforward integration of multiple sources of sensory input to achieve accurate vocal motor control during opera singing. Compared to non-singers, professional singers appear to integrate sensory inputs differently and to rely more on somatosensory feedback than on auditory feedback [33, 34].

During singing, afferent sensory information from the larynx, the articulating muscles, and the respiratory system reaches the somatosensory and insular areas via the posterior ventromedial thalamus [35].

This input is used in external sensory feedback by the primary auditory and somatosensory cortices to establish a relationship between motor efferents and sensory signals. Alongside this state-feedback system, a forward model involving premotor cortical areas predicts the state of the vocal effectors and their sensory consequences through efference copies of motor signals. Vocal training and expertise appear to enhance this forward system, which is updated by the feedback system if unexpected prediction errors or perturbations occur [33].

The role of the anterior insula in opera singing has been further confirmed in a further MRI study conducted on professionals engaged in traditional Chinese Pingju opera, an artistic genre integrating song, speech, and dance to express characters' emotions in a highly symbolic and conventional way [36] (Fig. 6.3).

Resting-state fMRI showed significantly lower spontaneous activity in the primary visual cortex and the dorsal visual pathway (a neuronal network involved in spatial processing) and higher activity in the anterior insula cortex. Furthermore, fMRI revealed an increased coherence of regional signal fluctuations in the left anterior insula, a region structurally and functionally connected with frontal areas involved in language processing, such as the frontal operculum and the prefrontal cortex [37].

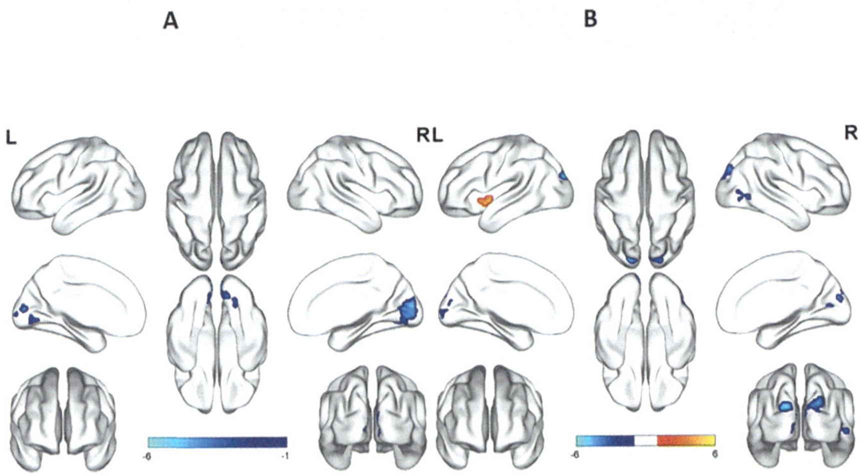

Fig. 6.3 Lower spontaneous regional brain activity in the visual cortex is shown in blue area (a) and higher brain activity in the anterior insula cortex is shown in orange area (b) in professional traditional Pingju opera actors, which might indicate superior performance in multidimensional professional skills such as music and face perception, dancing, and emotional representation. Zhang W, Zhao F, Qin W, Ma L. Altered Spontaneous Regional Brain Activity in the Insula and Visual Areas of Professional Traditional Chinese Pingju Opera Actors. Front Neurosci. 2018; 12: 450

These findings confirm the involvement of the anterior insula in vocal motor control. The activation of the anterior insula is associated with a feeling of well-being during singing, with some differences between females and males [38–40].

According to the "neural efficiency theory" [41], the decreased spontaneous activity observed in the cortical visual system might indicate that these opera singers and actors process visual information more effectively and at a lower threshold compared to laymen [36].

A further study from the same research group investigated neuroplasticity among professional Pingju opera actors who had undergone extensive training since childhood [42].

Participants were studied using arterial spin labeling magnetic resonance imaging to noninvasively evaluate resting-state cerebral blood flow at rest. Compared to untrained subjects, Pingju professionals showed regionally increased levels of resting-state cerebral blood flow. These alterations involved brain areas related to the processing of semantic aspects of language and empathy associated with dance or music (right middle temporal gyrus) [43, 44], regulation of emotional reactions to music and visual processing of facial images (right inferior temporal gyrus) [45, 46], multimodal sensory integration and audio-voice interactions (left temporal pole) [47], semantic and phonologic operations, and creativity (left inferior frontal gyrus) [48–50].

However, besides leading to adaptive changes in cerebral functions, long-term experience as an opera singer could also affect brain structure. One study investigated differences in gray matter volume between professional opera singers and non-singers using voxel-based morphometry. Compared to untrained subjects, opera singers had increased gray matter volume in cortical regions involved in motor control and feedback integration: the right ventral primary somatosensory and right rostral supramarginal gyrus, two adjacent areas co-activated during singing [51]; the right secondary somatosensory cortex, a region involved in the integration of auditory and somatosensory information [52]; and the right primary auditory cortex, a cortical area highly specialized in the processing and integration of pitch and timbre during singing [53].

An earlier beginning of vocal training was correlated with increased cortical volume in the ventral primary somatosensory region, but only in singers in whom training started beyond the age of 14 years, an age when speech development and control have matured. A growing body of research suggests that musical training has a beneficial impact on speech processing (e.g., hearing speech in noise and prosody perception) [54].

These findings are consistent with a prior tractography study showing an increased volume of the right arcuate fasciculus, a bundle of white matter fibers that connects frontotemporal, sensorimotor, and inferior parietal areas and cortical areas involved in music perception and vocal production and control [55].

These studies show that the somatosensory system is a neuronal network deeply involved in motor control and feedback integration that appears central in professional opera singing [56] (Table 6.1).

Table 6.1 Magnetic Resonance Imaging (fMRI) changes in Opera Singers Functional Brain during training vocal skills

Somato-motor and emotional features		
Automatic speech motor sequences into specialized motor programs (Increased gray matter and cortical thickness in auditory, visual, and motor regions)		
Cortical area	Increased activation of the primary somatosensory regions: somatotopic areas	Kinesthetic information and proprioceptive feedback of articulation and larynx
Cortical area	Fronto-parietal network (the inferior parietal lobe and bilateral dorsolateral prefrontal cortex)	Implicit and procedural motor memory
Cortical area	Left anterior insula (in connection with frontal and prefrontal areas)	Language processing
Subcortical area	Basal ganglia and thalamus and cerebellar networks	Implicit and procedural motor memory
Subcortical area	Increased volume of arcuate fasciculus (connects fronto-temporal, sensorimotor, and inferior parietal areas)	Music perception, and in vocal production and control
Emotional features		
Cortical area	Increased levels of cerebral blood flow right middle temporal gyrus	Processing of semantic aspects of language and empathy associated with dance or music
Cortical area	Right inferior temporal gyrus	Regulation of emotional reactions to music and visual processing of facial images
Cortical area	Left temporal pole	Multimodal sensory integration and audio-voice interactions
Cortical area	Left inferior frontal gyrus	Semantic and phonologic operations, and creativity
Cortical area	Anterior insula	Characters' emotions in a highly symbolic and conventional way: feeling of wellbeing during singing

6.4 Conclusions

Extensive vocal training in opera, a multimodal artistic genre based on the interaction of different sensorimotor components, can have profound effects on neuroplasticity. Spending years in training and practicing this highly specialized and multisensorial artistic genre can induce functional and structural adaptations in the human brain. These experience-dependent changes during singing mostly involve regional neuronal networks engaged in motor control of vocal activity and multimodal sensory integration, including auditory and sensorimotor regions and the insula. Further neuroimaging and neurophysiological studies conducted in larger samples and adopting complementary approaches (e.g., fMRI, tractography, and morphometry) are required to elucidate in greater detail the effects of opera singing on the structure and functions of the brain.

References

1. Matthews BR. Bravo! Neurology at the opera. In: Bogousslavsky J, Hennerici MG, Bäzner H, Bassetti C, editors. Neurological disorders in famous artists–Part 3. Basel: Karger; 2010. p. 119–29.
2. Hippocrates. Epidemics 2, 4–7. Cambridge: Harvard University Press; 1994.
3. Blair AM. The "problemata" as a natural philosophical genre. In: Grafton A, Siraisi N, editors. Natural particulars: nature and the disciplines in renaissance Europe. Cambridge: MIT Press; 1999. p. 171–204.
4. Pelosi F. Mind and Well-being in antiquity. In: Gouk P, Kennaway JG, Prins J, Thormälen W, editors. The Routledge companion to music, mind and Well-being. New York: Routledge; 2019. p. 19–31.
5. Borraccini M, Marinini M, Riva MA. Medieval interpretation of phonation through the divine comedy, by Dante Alighieri. Otolaryngol Head Neck Surg. 2022;1945998211073056:109. https://doi.org/10.1177/01945998211073056.
6. Maffei GC. Delle lettere, libri due, dove tra gli altri bellissimi pensieri di filosofia, e di medicina, v'è un discorso della voce e del modo d'apparare di cantar di Garganta, senza maestro, non più veduto, n'istampato, a cura di V. de Paoli da Limosano. Napoli: Raymundo Amato; 1562.
7. Huarte J. Examen de ingenious. Para la sciencias [...]. Baeça: Iuan Baptista De Montoya; 1575.
8. Martín-Araguz A, Bustamante-Martínez C. The examination of Men's Wits by Juan Huarte de san Juan, and the Dawn of the Neuropsychobiology of intelligence in the Spanish renaissance. Rev Neurol. 2004;38:1176–85.
9. Ulmus MA. Physiologia Barbae Humanae. Bononiae: apud Ioannem Baptistam Bellagambam; 1602.
10. Anatomiae JJ. Pragae, anno 1600 abs se solenniter administratae historia [...]. VVitebergae: excudebat Laurentius Seuberlich; 1601.
11. Riva A, Orrù B, Pirino A, Riva FT. Iulius Casserius (1552-1616): the self-made anatomist of Padua's golden age. Anat Rec. 2001;265:168–75.
12. Zytrowski A, Walocha J. Anatomical studies on larynx and voice production in historical perspective. Folia Med Cracov. 2020;60(3):85–98.
13. Solerti A. Gli albori del Melodramma. Milano-Palermo-Napoli: Sandron; 1904-1905.
14. Aristotele MG. Problemi di Fonazione e di Acustica. Napoli: Libreria Scientifica Editrice; 1962.
15. Hanning B. Of poetry and Music's power. Humanism and the creation of opera. Ann Arbor: UMI Research Press; 1980.
16. Wollock J. Communication disorders in renaissance Italy: an unreported case analysis by Hieronymus Mercurialis (1530-1606). J Commun Disord. 1990;23:1–30.
17. Palisca CV. Music and ideas in the sixteenth and seventeenth centuries. In: Mathiesen TJ, editor. Studies in the history of music theory and literature. Champaign: University of Illinois Press; 2017.
18. Gouk P. Melancholy, music and the passion in English culture around 1600. In: Gouk P, Kennaway JG, Prins J, Thormälen W, editors. The Routledge companion to music, mind and Well-being. New York: Routledge; 2019. p. 63–76.
19. Browne R. Medicina musica. London: Printed for J. Cooke; 1729.
20. Brocklesby R. Reflections on antient and modern musick, with the application to the cure of diseases. London: Printed for M. Cooper; 1749.
21. MacDonalds RAR, Kreutz G, Mitchell L. Music, health, and wellbeing. New York: Oxford University Press; 2012.
22. Puderbaugh M, Emmady PD. Neuroplasticity. In: StatPearls [internet]. Treasure Island: StatPearls Publishing; 2022.
23. Jäncke L. The plastic human brain. Restor Neurol Neurosci. 2009;27:521–38.
24. Munte TF, Altenmuller E, Jancke L. The musician's brain as a model of neuroplasticity. Nat Rev. 2002;3:473–8.

25. Bermudez P, Zatorre RJ. Differences in gray matter between musicians and nonmusicians. Ann N Y Acad Sci. 2005;1060:395–9.
26. Bermudez P, Lerch JP, Evans AC, Zatorre RJ. Neuroanatomical correlates of musicianship as revealed by cortical thickness and voxel-based morphometry. Cereb Cortex. 2009;19:1583–96.
27. Gaser C, Schlaug G. Brain structures differ between musicians and non-musicians. J Neurosci. 2003;23:9240–5.
28. Smith A, Zelaznik HN. Development of functional synergies for speech motor coordination in childhood and adolescence. Dev Psychobiol. 2004;45:22–33.
29. Kleber B, Veit R, Birbaumer N, Gruzelier J, Lotze M. The brain of opera singers: experience-dependent changes in functional activation. Cereb Cortex. 2010;20:1144–52.
30. Huettel SA. Event-related fMRI in cognition. NeuroImage. 2012;62(2):1152–6.
31. Lee MH, Smyser CD, Shimony JS. Resting-state fMRI: a review of methods and clinical applications. AJNR Am J Neuroradiol. 2013;34(10):1866–72.
32. Logothetis NK. What we can do and what we cannot do with fMRI. Nature. 2008;453(7197):869–78.
33. Kleber B, Zeitouni AG, Friberg A, Zatorre RJ. Experience-dependent modulation of feedback integration during singing: role of the right anterior insula. J Neurosci. 2013;33(14):6070–80.
34. Kleber B, Friberg A, Zeitouni A, Zatorre R. Experience-dependent modulation of right anterior insula and sensorimotor regions as a function of noise-masked auditory feedback in singers and nonsingers. NeuroImage. 2017;147:97–110.
35. Jürgens U. Neural pathways underlying vocal control. Neurosci Biobehav Rev. 2002;26(2):235–58.
36. Zhang W, Zhao F, Qin W, Ma L. Altered spontaneous regional brain activity in the insula and visual areas of professional traditional Chinese Pingju opera actors. Front Neurosci. 2018;12:450.
37. Augustine JR. Circuitry and functional aspects of the insular lobe in primates including humans. Brain Res Brain Res Rev. 1996;22:229–44.
38. Ascenso S, Perkins R, Williamon A. Resounding meaning: a PERMA wellbeing profile of classical musicians. Front Psych. 2018;8:1895.
39. Craig AD. How do you feel–now? The anterior insula and human awareness. Nat Rev Neurosci. 2009;10:59–70.
40. Demaree HA, Everhart DE, Youngstrom EA, Harrison DW. Brain lateralization of emotional processing: historical roots and a future incorporating dominance. Behav Cogn Neurosci Rev. 2005;4:3–20.
41. Del Percio C, Babiloni C, Bertollo M, Marzano N, Iacoboni M, Infarinato F, et al. Visuo-attentional and sensorimotor alpha rhythms are related to visuo-motor performance in athletes. Hum Brain Mapp. 2009;30:3527–40. https://doi.org/10.1002/hbm.20776.
42. Zhao F, Zhang W, Zhu D, Wang X, Qin W, Liu F. Long-term Pingju opera training induces plasticity changes in cerebral blood flow: an arterial spin labelling MRI study. Neuroscience. 2020;436:27–33.
43. Hickok G, Poeppel D. The cortical organization of speech processing. Nat Rev Neurosci. 2007;8:393–402.
44. Gujing L, Hui H, Xin L, Lirong Z, Yutong Y, Guofeng Y, Jing L, Shulin Z, et al. Increased insular connectivity and enhanced empathic ability associated with dance/music training. Neural Plast. 2019;9693109:1.
45. Baumgartner T, Lutz K, Schmidt CF, Jäncke L. The emotional power of music: how music enhances the feeling of affective pictures. Brain Res. 2006;1075:151–64.
46. Eldar E, Ganor O, Admon R, Bleich A, Hendler T. Feeling the real world: limbic response to music depends on related content. Cereb Cortex. 2007;17:2828–40.
47. Ohki T, Gunji A, Takei Y, Takahashi H, Kaneko Y, Kita Y, Hironaga N, Tobimatsu S, et al. Neural oscillations in the temporal pole for a temporally congruent audio-visual speech detection task. Sci Rep. 2016;6:37973.

48. Tanaka K, Ohta S, Kinno R, Sakai KL. Activation changes of the left inferior frontal gyrus for the factors of construction and scrambling in a sentence. Proc Jpn Acad Ser B Phys Biol Sci. 2017;93:511–22.
49. Giustolisi B, Vergallito A, Cecchetto C, Varoli E, Romero Lauro LJ. Anodal transcranial direct current stimulation over left inferior frontal gyrus enhances sentence comprehension. Brain Lang. 2018;176:36–41.
50. Okada K, Matchin W, Hickok G. Phonological feature repetition suppression in the left inferior frontal gyrus. J Cogn Neurosci. 2018;30:1549–57.
51. Grabski K, Lamalle L, Vilain C, Schwartz JL, Vallee N, Tropres I, Baciu M, Le Bas JF, Sato M. Functional MRI assessment of orofacial articulators: neural correlates of lip, jaw, larynx, and tongue movements. Hum Brain Mapp. 2012;33:2306–21.
52. Eickhoff SB, Grefkes C, Zilles K, Fink GR. The somatotopic organization of cytoarchitectonic areas on the human parietal operculum. Cereb Cortex. 2007;17:1800–11.
53. Özdemir E, Norton A, Schlaug G. Shared and distinct neural correlates of singing and speaking. NeuroImage. 2006;33:628–35.
54. Schloneger MJ. Graduate student voice use and vocal efficiency in an opera rehearsal week: a case study. J Voice. 2011;25(6):e265–73.
55. Halwani GF, Loui P, Rüber T, Schlaug G. Effects of practice and experience on the arcuate fasciculus: comparing singers, instrumentalists, and non-musicians. Front Psychol. 2011;2:156.
56. Proverbio AM. Il cervello del cantante. In: Proverbio AM, editor. Neuroscienze cognitive della musica. Bologna: Zanichelli Editore; 2021. p. 19–33.

Part III

ACT III Diseases that Affect Workers in the Opera World

Operatic Inspiration, Genius and Madness

7

Roberto Mazzagatti

7.1 Introduction

The relationship between genius and madness has ancient origins. Indeed, the theory was contemplated by Aristotle (384–322 BC) in his collection Problemata (Book XXX), in the passage in which the philosopher states that a great mind cannot exist without a touch of madness. Since Aristotle, other classical and medieval scholars, when discussing talent, suggested a link with 'black bile' or 'melancholy'—a term used in Hippocratic and Galenic doctrines to identify mental illness. The juxtaposition of genius and psychopathy later became a topos of Romantic science and took root in mass culture, especially through the circulation of Cesare Lombroso's (1835–1909) seminal work, Genio e follia (1864), published in English as The Man of Genius (1889). In this work, the scientist took a radical stance, arguing that genius should be regarded as a trait intrinsically linked to insanity. Henceforth, an increasing number of scholars in the twentieth century addressed the connection between creative potential and psychoticism with a largely empirical approach, essentially consolidating the research inherited from nineteenth-century neuropsychology [1].

In the artistic field—but now also in the field of scientific research and, more generally, in intellectual professions—imagination is considered a necessary and essential quality. This way of thinking is referred to as 'overinclusive thinking' and involves the mental processing of loose (unusual) associations, often accompanied by eccentric attitudes. The biographies of artists, writers and musicians are full of this kind of bizarre and delusional behaviour, which implicitly seems to confirm the crossover between creativity and mental illness in the past.

R. Mazzagatti (✉)
School of Medicine and Surgery, University of Milano-Bicocca, Monza, Italy
e-mail: roberto.mazzagatti@unimib.it

L. Lorusso et al. (eds.), *Effects of Opera Music from Brain to Body*, Neurocultural Health and Wellbeing,
https://doi.org/10.1007/978-3-031-34769-6_7

The myth of the mad genius has maintained the appeal and fascination it has exerted since the dawn of the history of thought; as theoretical and experimental tools have become more sophisticated, science too has sought to understand whether a correlation can indeed be established between mental and nervous disorders and a particular talent or above-average intelligence. The aim of modern research is also to understand whether highly creative people are more likely to suffer from neurological diseases or develop mental illnesses, in order to develop appropriate preventive measures.

The results of this scientific research, it should be noted, refute the existence of a causal relationship between symptoms of mental disorder and creative work at the highest level, but do identify a correlation between the two conditions. At present, however, the neurocerebral mechanism underlying this phenomenon remains to be elucidated in a number of respects [2–4].

7.2 Writers, Artists with Creativity and Madness

If we look at the history of literature, philosophy, art and entertainment, the theme of madness is extensive. Numerous figures who had an original vision of reality, or true geniuses in their field, straddled this line with psychopathology.

Some of the most famous names include John Keats (1795–1821), Charles Baudelaire (1821–1867), Leo Tolstoy (1828–1910), Édouard Manet (1832–1883), Friedrich Nietzsche (1844–1900), Paul Gauguin (1848–1903), Guy de Maupassant (1850–1893), Vincent van Gogh (1853–1890), Oscar Wilde (1854–1900), Mikhail Vrubel (1856–1910), Theo van Gogh (1857–1891), Henri de Tolosa-Lautrec (1864–1901), Eugen Sandow (1867–1925), Kostas Karyotakis (1896–1928), Sylvia Plath (1932–1963) and John Nash (1928–2015). This list, albeit partial, nevertheless spans many areas of human production, including mathematical reasoning. This shows that in any creative occupation, genius is frequently accompanied by uncommon behaviours—such as repetitive and ritual actions—and often by actual diagnoses of obsessions, mood disorders, psychosis, schizotypal personality disorder and even schizophrenia.

In The Genesis of Artistic Creativity: Asperger's Syndrome and the Arts, Michael Fitzgerald, former Professor of Child and Adolescent Psychiatry at Trinity College Dublin and a clinician with extensive experience in the diagnosis of Asperger's syndrome (a form of high-functioning autism), recounts the stories of a number of cultural and artistic figures who, in his opinion, manifested symptoms that could be correlated with autism. Fitzgerald draws on these retrospective diagnoses to suggest, firstly, that creative ability has a bearing on autism and, secondly, that creativity is a mental faculty inherent in psychiatric disorders. In other words, talent has a purely internal cause, while external variables, such as education and willpower, are not sufficient to determine the performance of a high achiever [5].

The idea that mental illness is linked with the creative experience has also been confirmed by researchers at the Karolinska Institutet in Stockholm, who have monitored over a million people living in Sweden for more than 40 years [6]. This study had the advantage of overcoming the limitations of previous empirical investigations, which lacked scientific reliability due to their over-dependence on biographical data and their failure to include a sufficiently broad sample of patients.

The results of this research indicate that in the Swedish population, the incidence of schizophrenia and bipolar disorder is higher among musicians, visual artists (painters and photographers), scientists and dancers. The risk of developing these conditions was also found to be shared by relatives of the patients in the study cohort. Furthermore, and compared with the general population, the above-mentioned categories demonstrate increased risk for schizophrenia, anxiety and depression, whereas in relation to other psychopathologies there is no significant deviation from the average.

In summary, the Swedish study substantiates with sufficient breadth and scientific reliability the existence of a relationship between a career in the arts and a psychiatric diagnosis. The creative mind is, therefore—in light of the science and the historical evidence—more prone to suffer from schizophrenia, bipolar disorder, depression, anxiety, addictions or suicide, and this predisposition clearly reflects a genetic cause, since there is a firm correlation between artistic professions and a family history of psychiatric disorders.

Madness has been a classic topos in literature in all periods (suffice it to mention the authors Homer and Ludovico Ariosto to recall famous writings), but it is also a recurring theme in opera music. Looking at the history of music, mental illnesses were not only elements of the plot or of the characterization of operatic characters—see Gaetano Donizetti's *Lucia di Lammermoor* and *Anna Bolena* (1797–1848), Igor Stravinsky's *Career of a Libertine* (1882–1971) or Alban Berg's *Wozzeck* (1885–1935)—but were often conditions experienced by the composers themselves who, despite their troubled psyche, succeeded in creating sublime works, perhaps even deriving positive effects for their creativity.

In view of the particular convergence between creative disposition and mental health, when exploring lyrical inspiration and medical issues in music, it is worth recalling the well-known cases in which the burden of mental illness is to some extent offset by the advantage of greater artistic genius.

It is a historically established fact that many of the most eminent figures of the Neoclassical and Romantic musical scene exhibited an intolerance for reality—a probable symptom of underlying mental health issues—and some of them spent the final years of their lives in an asylum.

Approaching the issue from a historical perspective, however, a premise must be noted: between the sixteenth century and the first half of the twentieth century, psychoses appear to have had a greater epidemiological impact than today. One of the many reasons for this is the widespread prevalence of syphilis in historical populations, as the disease, which was incurable at the time, commonly resulted in general paresis: a condition involving behavioural disorganization, delusions, paranoia and other neurological deficits. If we exclude, firstly, alcohol and drug abuse and, secondly, neuroses—which were nevertheless over-represented among artists, who used them as a functional response to talent management and emotional regulation—syphilis is indeed one of the diseases that has had the greatest influence on the work of composers over the last 500 years [7].

The relationship between syphilis and musical genius is particularly apparent in the more mature stage of the work of composers, which is evidently linked to the long-term progression of this infection. In the sixteenth century, the fulminant form of syphilis was rampant, but it later evolved into a slowly developing disease with three progressive stages. Neuropsychiatric complications, with their varied clinical manifestations, usually began to appear several years or even a decade or more after infection and in a progressive way. However, syphilis cannot always be attributed to specific figures: there have been few cases in which a patient—especially a public figure—has admitted to suffering from the disease. Although there is evidence that syphilis was endorsed in certain literary circles by this time—Gustave Flaubert (1821–1880) and Charles Baudelaire (1821–1867) apparently flaunted it as a badge of distinction and value—in polite society, the disease was still considered dishonourable as it was linked to abject sexuality and sin. Moreover, this affliction disfigured the body, could impair the mental faculties and lead slowly but inexorably to death. All of these aspects made this disorder even more psychologically unacceptable [8].

Despite the propaganda of the libertines, who wanted to exalt the venereal infection as a symbol of free thought and the revolution in sexual mores, syphilis continued to be perceived as a sin until well into the nineteenth century, meaning that patients almost always concealed the true nature of the disease. It was also not uncommon for the families of the infected, including well-known musicians, to circulate stories covering up the decline and death of their relatives.

7.3 Musicians, Composers and Syphilis

Based on what can be reconstructed, at least eight famous composers probably contracted the disease and suffered neurological complications to varying degrees. These are Niccolò Paganini (1782–1840), Franz Schubert (1797–1828), Gaetano Donizetti (1797–1848), Robert Alexander Schumann (1810–1856), Bedřich Smetana (1824–1884), Hugo Wolf (1860–1903), Frederick Delius (1862–1934) and Scott Joplin (1867/1868–1917) [9].

It has also been suggested that Ludwig van Beethoven (1770–1827) suffered from syphilis and may have become deaf as a result of tabes dorsalis. In the same way that the syphilitic nature of Beethoven's deafness remains a matter of conjecture, the origin of Robert Schumann's personality change, as we shall see, also remains uncertain. The symptoms experienced by Hugo Wolf, Frederick Delius and Scott Joplin, on the other hand, can be attributed with a degree of certainty to *Treponema pallidum* [7].

However, if many opera composers from the golden age of classical music were affected by the infection, this is perhaps no coincidence—and indeed, we might be forgiven for wondering whether the disease itself was the source of the creative impetus that shaped famous musical masterpieces [10]. In their final years, tertiary syphilis patients often experience a gradual onset of dementia, paranoia and hallucinations, including auditory ones—sometimes transient, sometimes permanent—but which, in the case of composers, may have contributed to inspiring the original ambience of their most famous compositions. In addition to the aforementioned effects of the infection, there may also be evidence of mental disorders of an iatrogenic nature in the later works of musicians with the disease. The mercury therapies that were attempted at the time against the disease are likely to have aggravated rather than alleviated the neurological manifestations, as such treatments were highly toxic to the nervous system. The palaeopathological examination of Bedřich Smetana's remains, for example, revealed not only a venereal infection but also high concentrations of mercury. These examinations thus support the hypothesis of medical historians concerning the senseless administration of the substance in large doses in times prior to the introduction of antibiotic therapy.

In the case of the Czech composer, it should be added that autopsy examinations carried out immediately after his death in 1884 in a psychiatric institution in Prague had already established the presence of syphilis. Unfortunately, Smetana began to suffer from auditory hallucinations at an early age, and by the time he turned 40, his ear problems had worsened, leading first to tinnitus and then to hearing loss. In the space of a few years, Smetana's mental faculties deteriorated. Despite treatment by the most eminent specialists in Paris, he sank inexorably into dementia and developed paranoid tendencies and violent reactions. In his semi-infirm state, the musician ranted about visits he had never received, wrote letters from the asylum to imaginary people and rambled incoherently in a mixture of Czech and German [10].

The composer Gaetano Donizetti also spent his last years in an asylum because of mental disorders brought on by meningovascular syphilis. Donizetti was only released from the psychiatric hospital when he was close to death and bid farewell to his native Bergamo, where he died in the patrician residence of Palazzo Basoni Scotti.

The musician's correspondence reveals the impact that the illness had on his character. In the last years of his career, Donizetti seemed to change: he became more suspicious, his mood worsened and he seemed increasingly gripped by grim paranoia. With feelings of persecution, he developed a stubborn distrust of others and gradually withdrew into himself. Philippe Ricord, a specialist in venereal diseases who was very much in vogue among politicians and artists in mid-nineteenth-century Paris, treated him to no avail [11].

In the throes of mental derangement, by then irreversible, Donizetti's family had to induce him to believe he was spending a holiday in a hotel in order to have him admitted to the asylum. The artist must have suffered a great deal during the last years of his life, as the autopsy revealed brain damage that undoubtedly caused him to be tormented by quadriplegia, convulsions and terrible migraines [7].

Donizetti should also be mentioned here for having staged two of the most evocative and powerful representations of mental disorder, namely, *Anna Bolena*'s hallucinatory imprisonment in the opera of the same name and the psychotic breakdown of the protagonist in *Lucia di Lammermoor* (Fig. 7.1).

Fig. 7.1 Giuditta Pasta (1797–1865) in the role of Anne Boleyn (nineteenth century) by composer Gaetano Donizetti (1797–1848). Picture of the Russian painter Karl Pavlovich Bryullov (1799–1852)

In these scenes, fraught with unease and obsession, the composer succeeded in translating the distorting reality of mental imbalance into music and—perhaps influenced by the early neurological symptoms of his illness—in reproducing on stage an incisive, convincing and unforgettable image of the loss of sense and delirium that precedes madness.

While Donizetti's experience of illness has offered scholars a key to understanding his singular ability to capture the inner dimension and outward behaviour of a psyche gripped by madness, this aspect, in other respects, cannot but generate in audiences of his operas a feeling of disorientation and a certain apprehension. It is indeed distressing to recall that the author of these unparalleled representations of madness was himself consumed by insanity and spent his last years in isolation in the asylum in Ivry-sur-Seine. The epilogue is tragically reminiscent of the ill-fated Anne Boleyn, Henry VIII's desperate consort confined in captivity in the Tower of London [12].

To conclude this discussion on the historiography of syphilis, it is also worth mentioning the research into the effects of the disease on the work of great rulers, religious, military and mystical figures. The aim is equally explicit and involves combining the pathological element with the individual's behaviour. At this juncture, however, the aim is to propose new theories regarding certain political responsibilities, moral choices and behaviour.

A prime example is the likely centrality of neurosyphilis in the military tactics of the last battles fought by Napoleon Bonaparte (1769–1821). During the last years of the emperor's life, it is thought that he experienced tertiary symptoms of the disease, including the emergence of mental problems and the persistence of troublesome urinary disorders. According to accounts of the Russian campaign, the commander's mental decline was indeed sensed by his staff, who described Napoleon as incapable of making sensible decisions, plunged into a state of exaltation consisting at times of a hallucinatory interpretation of reality. It goes without saying that these effects cannot have been in any way beneficial as paradoxically they may be for those working in the artistic field. Indeed, in the case of the legendary general, madness did not enrich his genius. Quite the opposite. From this perspective, syphilis contributed fatally to his final defeat, consuming his strategic lucidity and leaving Napoleon with a volatile and hesitant personality [8].

7.4 Neuropsychiatric Disorders in Musicians and Composers

As we explore the field of diseases that may have a bearing on musical ability, apart from neurosyphilis, we also encounter melancholy states and neurological disorders. A wide range of symptoms can spill over into musical writing and alter the perception of rhythm, harmonic lines and polyphony to such an extent that they become an element that is difficult to separate from the author's style. The depression suffered by Giuseppe Verdi (1813–1901) and the early-onset dementia that affected Maurice Ravel (1875–1937) are worth mentioning here.

Giuseppe Verdi experienced the symptoms of a grim affliction during the last decades of his life. The great opera composer may have been able to draw on this experience of suffering as well as on his condition as a patient. The representation of mental suffering in Verdi's work indeed seems to be influenced by this personal experience, but it is also likely that Verdi's psychiatrist and caregiver, Cesare Vigna (1819–1892), contributed to the depiction of the condition in his work.

The relationship between the composer and Vigna began through Verdi's second wife, Giuseppina Strepponi (1815–1897). Indeed, it was Strepponi who had approached the psychiatrist, who was director of the women's asylum in San Clemente (Venice), after the composer had fallen into a state of depression.

Over time, a friendship developed between the composer and the doctor, resulting in a professional partnership evidenced by regular correspondence. Neuropsychiatric disorders are represented in the plots of Verdi's operas: delirium or acute psychosis in *Nabucco* (1842), madness in *Giovanna d'Arco* (1845), somnambulism in *Macbeth* (1847) and insane jealousy in *Otello* (1887). They are, however, portrayed with lyrical substance, which does not conform to the romantic stereotype and perhaps also allows us to glimpse the contribution of an experienced specialist [13] (Fig. 7.2).

Fig. 7.2 Title page from the variant first edition vocal score of Giuseppe Verdi's Giovanna d'Arco (1846), designer Girolamo Magnani (1815–1889)

Moreover, Vigna had experimented with the effects of music in the treatment of mental disorders and had published the results of his studies in the *Gazzetta musicale di Milano* between 1852 and 1856. The doctor wanted to shed light on the neurophysiological processes triggered by music and to apply them in the clinic of nervous and mental diseases. Vigna was aware of the research initiated in France in the years 1820–1830 by Jean-Étienne Dominique Esquirol (1772–1840) at the Parisian hospital of La Salpêtrière and it was in this context that he pursued his research in the psychiatric hospital he directed. At San Clemente at the time, a room in the facility had been specially set aside for music therapy, to which patients had free access.

It has been argued that neurological pathology can have favourable consequences, but it is very difficult to establish retrospectively where personal innovation ends and the effect of an organic alteration takes over. Especially when there are doubts as to the aetiology of the disorders, as is the case for patients living at a time when reliable biological methods of diagnosis were not yet available.

Such was the case of the Frenchman Maurice Ravel (1875–1937) who, at the age of 52, at the peak of his career, began to show signs of a neurological deterioration that altered his musical sensitivity and distorted his compositional style. The root of Ravel's illness cannot be clearly deciphered, though some scholars have interpreted it as Alzheimer's disease [14, 15].

According to research carried out by François Boller, Ravel's condition appears to be the result of damage primarily to the left cerebral hemisphere, which progressively degenerated into two major deficits: loss of verbal functions (aphasia) and loss of movement control (apraxia) [16].

Although Ravel's illness—which affected the speech centres of his brain and led to corticobasal degeneration—compromised his mobility and thus his ability to write and compose piano pieces over a period of 5 years, it did not affect his musical ability in the strictest sense. Musical ability, it should be noted, is not only localized in the left hemisphere, but seems to extend to both hemispheres.

On the basis of functional considerations, Boller and his colleagues claim that in the creation of Ravel's last two compositions, the dominance of the preserved cerebral hemisphere, the right one, and the weakening of the diseased hemisphere, the left one, is evident. In *Boléro* (1928) and the *Piano Concerto for the Left Hand* (1929–1930), the orchestral timbres developed by the right hemisphere predominate, while the melodic complexity of the left brain is secondary. Ravel's final works thus demonstrate how the illness can randomly affect the creative process and distort the surviving musical faculties to produce an exhilarating, unusual and enjoyable score.

While Ravel's unique melodic inventiveness may have been fuelled by an intermediate state between normality and cognitive disability, Niccolò Paganini's prodigious mastery of the violin seems to be attributable to a rare connective tissue disease [17].

It is well known that Paganini suffered from numerous health problems: in his early teens he contracted measles—a disease that later deteriorated into severe encephalitis—and later fell ill with tuberculosis and syphilis. However, according to several scholars, his virtuosity can be explained by a genetic anomaly known as Marfan syndrome.

The violinist's correspondence and medical records seem to support this interpretation. Moreover, the theory implicating a hereditary disease has recently been confirmed by iconodiagnostic analyses of paintings of Paganini and on a cast of his right hand made the day after his death [18].

The syndrome in question would therefore explain his emaciated appearance, skeletal body and long, slender and extremely flexible fingers that recur in portraits of the Genoese musician. In particular, arachnodactyly is believed to have been the anatomical basis for his extraordinary virtuosity with the violin. It would, however, be unfair to attribute the secret of Paganini's ability solely to skills acquired through the illness from which he suffered. Certainly, with his disproportionately long fingers, he was able to effortlessly reach any point on the instrument with rare dexterity and rapidity, but apart from the 'genetic advantage', it should not be overlooked that the maestro constantly underwent a gruelling routine of daily practice (Fig. 7.3).

Fig. 7.3 The violinist Niccolò Paganini (1782–18409) by Georg Friedrich Kersting (1785–1845), picture of 1829, Staatliche Kunstsmmlungen Dresden

The success of Paganini, who is celebrated as the greatest violinist of all time, must therefore be considered a fortuitous and fortunate combination of several factors: firstly, his possession of a physical requirement (arachnodactyly) brought on by Marfan syndrome and which made it difficult for other instrumentalists to achieve the same ability and, secondly, his personal aptitude for willpower and perfectionism, which enabled him to reach a playing speed that was unattainable for other violinists.

Returning to stories about opera composers that suffered from psychosis, a particularly interesting case is that of Robert Alexander Schumann's (1810–1856) bipolar disorder. Our knowledge of Schumann's life appears to contradict the stereotypical view of the pairing of genius and madness.

Scholars have long been in doubt as to the clinical definition of the illness, which profoundly affected the composer's work for a long time. Schumann suffered from paracusis, and the musician had revealed to his wife and friends that he heard music: at first he perceived a divine melody but later these sound hallucinations degenerated into an eerie noise or an uninterrupted, high-pitched A note. The diagnosis of syphilis made by the doctors who treated Schumann at the time was initially accepted by scholars. Subsequently, meningioma, a form of brain cancer that can lead to confusion, amnesia and hearing problems, was suggested. At present, researchers seem to be fairly unanimous in recognising Schumann's manic-depressive psychosis, a disorder that seems to have some basis in genetic transmission, complicated by syphilis contracted years earlier and by then in the encephalitic stage. The German musician's death has also been linked to the lethal effects of the mercurial drugs used in the treatment of the disease [19].

It is conceivable that the musician struggled with the hallucinations, perhaps even initially managing to master them and use them in the creative process, but at some point these phenomena must have become uncontrollable, relentlessly damaging his work. By 1850, Robert Schumann was a man exasperated by his hallucinatory delirium, so exhausted that he was unable to keep his engagement as general director of music in Düsseldorf. His later years evoke disorganization, withdrawal and profound dissociation rather than the decadent myth of recklessness and genius. During his stay in Düsseldorf, he is said to have been very disturbed by his ailments and to have remained absorbed for hours as if in a state of dissociation from reality.

It should be noted that Schumann's already fragile mental balance was perhaps aggravated by certain circumstances he had to face during his life, first and foremost, the regret of having to give up his dream of pursuing a concert career. His aspiration to become a great pianist was probably undermined by an occupational neurosis that struck him in 1831–1832. The disorder is referred to as the musician's dystonia, as it is mainly suffered by pianists and violinists and is characterized by involuntary spasms due to continuous strain on the finger flexors and especially when performing rapid movements with small muscles.

Schumann had developed this problem in his right hand and unfortunately attempts to remedy it with the aid of a dactylion (a device that stretched the fingers then in vogue among pianists) caused him permanent damage. The musician irrevocably lost the use of his ring finger and the possibility of pursuing a career as a pianist [19].

A further blow to his stability came with the loss of his loving mother and two of his brothers. Schumann struggled to recover from these bereavements and perhaps began to drown his sorrows in alcohol. The composer's life came to an end in 1854. After attempting to take his own life by jumping into the Rhine, Robert Schumann was admitted to the asylum in Endenich, just outside Bonn, where he died in 1856 after an almost complete mental breakdown, relieved only by occasional visits from his wife Clara.

In conclusion, the stories of the composers outlined so far seem to point to a dual effect of these disorders on musical ability: on the one hand, a mental disorder can enrich talent and facilitate the achievement of a work of art, but on the other hand, it can also make a person more vulnerable and erratic in their work.

Despite its ambiguous nature, the myth of the mad genius enshrined in popular culture by Lombrosian theory over the last 150 years continues to fascinate scientists. In recent years, new investigations have focused on autism, attributing the disorder to several well-known figures from classical music, painting and literature. To name a few, Wolfgang Amadeus Mozart (1756–1791), Ludwig van Beethoven, Vincent van Gogh (1853–1890), Andy Warhol (1928–1987), Hans Christian Andersen (1805–1875) or George Orwell (1903–1950) may have achieved success at the expense of serious personal problems.

Psychosis and autism spectrum disorders unfortunately have serious consequences for those affected. The composers in question were often shy or rebellious, hesitant and capable of violent behaviour, as well as being incapable of leading a normal life, i.e. providing for their daily needs and fulfilling their social and professional obligations [20].

In the history of twentieth-century music, one case of synergy between talent and madness is that of the Canadian pianist Glenn Gould (1932–1982) [21]. Widely considered by critics to be a musician with an extraordinary technique and an outstanding performer, Gould was also severely attacked as a person because of his eccentric and sometimes incomprehensible attitudes and as a musician because of his unusual and daring performance of the classical repertoire, which he reinterpreted in an unorthodox manner, one might say, that did not adhere to its original character.

Gould suffered from stereotypies and observed a strict routine: he only played sitting on the chair his father had built for him and in the theatre he demanded a precise (very hot) temperature to perform. As well as being alarmingly superstitious, he was extremely fearful of the cold and detested contact with other people and there were many reports of inappropriate behaviour by colleagues and associates. After retiring from the stage in 1964 at the age of 32, Gould lived in his own home and had only sporadic contact with the outside world. Although anecdotes about him do hint at markedly schizoid traits, Gould himself undertook to refute this slander by publishing his defence in a book entitled *No, I am not an eccentric*. Today, however, scholars are inclined to consider Glenn Gould's eccentricities as caused by obsessive-compulsive disorder and Asperger's syndrome, so his extraordinary genius is further indicative of the link between a pinch of healthy madness and musical excellence.

7.5 Conclusions

Ultimately, perhaps observing brain degeneration from the point of view of the artistic value it can generate—as well as helping us to clarify certain aspects of the functioning of the mind and potentially being of some use in the treatment of sufferers—could help us to reject the age-old distrust of this kind of patient, centuries ago even perceived as demon-possessed, and to accept this human condition as one of the possible expressions of life. A way of being in the world, mysterious minds that could even hold extraordinary gifts, because 'madmen' are also extraordinary artists among men.

References

1. Eysenck HJ. Genius. The natural history of creativity. Cambridge: Cambridge University Press; 2008.
2. Simonton DK. Can creative productivity be both positively and negatively correlated with psychopathology? Yes! Front Psychol. 2014;5:455.
3. Batey M, Furnham A. Creativity, intelligence, and personality: a critical review of the scattered literature. Genet Soc Gen Psychol Monogr. 2006;132(4):355–429.
4. Barron F. Originality in relation to personality and intellect. J Pers. 1957;25(6):730–42.
5. Fitzgerald M. The genesis of artistic creativity: Asperger's syndrome and the arts. London and Philadelphia: Jessica Kingsley Publishers; 2005.
6. Kyaga S, Landén M, Boman M, Hultman CM, Långström N, Lichtenstein P. Mental illness, suicide and creativity: 40-year prospective total population study. J Psychiatr Res. 2013;47(1):83–90.
7. Pedro MKF, Germiniani FMB, Teive HAG. Neurosyphilis and classical music: the great composers and "the great imitator". Arq Neuropsiquiatr. 2018;76(11):791–4.
8. Snowden FM. Storia delle epidemie. Dalla Morte Nera al Covid-19. Gorizia: Leg edizioni; 2020.
9. Sifilide GP. Scienza, storia, costume, letteratura. Arghilla (RG): Besa Muci; 2021.
10. Rempelakos L, Poulakou-Rebelakou E, Tsiamis C, Rempelakos A. Syphilis impact on late works of classical music composers. J Urol. 2014;191(4):29–36.
11. Renner C. A propos de Philippe Ricord [about Philippe Ricord]. Hist Sci Med. 2008;42(4):365–70.
12. Peschel E, Peschel R. Donizetti and the music of mental derangement: Anna Bolena, Lucia di Lammermoor, and the composer's neurobiological illness. Yale J Biol Med. 1992;65(3):189–200.
13. Riva MA, Lorusso L, Sironi VA. Cesare Vigna (1819-1892). J Neurol. 2014;261(2):449–50.
14. Baeck E. Was Maurice Ravel's illness a corticobasal degeneration? Clin Neurol Neurosurg. 1996;98(1):57–61.
15. Henson RA. Maurice Ravel's illness: a tragedy of lost creativity. Br Med J (Clin Res Ed). 1988;296(6636):1585–8.
16. Amaducci L, Grassi E, Boller F. Maurice ravel and right-hemisphere musical creativity: influence of disease on his last musical works? Eur J Neurol. 2002;9(1):75–82.

17. Schoenfeld MR. Nicolo Paganini. Musical magician and Marfan mutant? JAMA. 1978;239(1):40–2.
18. Pedrazzini A, Martelli A, Tocco S. Niccolò Paganini: the hands of a genius. Acta Biomed. 2015;86(1):27–31.
19. Sacks O. Musicophilia. Tales of music and the brain. New York: Random House Abridged Edition; 2007.
20. Proverbio AM. Genio musicale e "follia". In: Proverbio AM, editor. Percezione e creazione musicale. Bologna: Zanichelli Editore; 2022. p. 175–92.
21. Newmark J. Neurological problems of famous musicians: the classical genre. J Child Neurol. 2009;24(8):1043–50.

8 Work and Disease in Italian Operas

Michele A. Riva

8.1 Opera, Work and Disease in Italy

Operas, as an expression of their society, were not only popular shows, the audience being also composed of upper-class and more educated people. Thus, they could be considered as a crossing point between the popular and the intellectual world [1]. As an art form, opera cannot but be embedded in the social context it comes out of. Hence, operas may provide us with valuable information on the society of that time, on the conception and meaning of the work, on the workers' health conditions and on the progressive acknowledgment of work-related diseases in medical and popular opinion.

The relationship between operas, work conditions and occupational diseases may be well shown in Italy and particularly in the city of Milan. During the nineteenth century, Milan was also the most dynamic metropolitan city in Italy and the main centre of the Italian Industrial Revolution. A large-scale cotton industry developed around 1840, particularly in and around Milan. Steel plants, chemical industries, machine tool companies and cement works emerged in the area at the end of the century. The worsening of working conditions soon led to an increase in work accidents and to the creation of workers' compensation insurances in Milan, protecting workers against physical and economic damages deriving from accidents caused by the working activity. Occupational diseases were diagnosed and treated at the *Clinica del Lavoro*, the first occupational health institute in the world, founded in Milan by Luigi Devoto (1864–1936) in 1902 [2, 3].

M. A. Riva (✉)
School of Medicine and Surgery, University of Milano-Bicocca, Monza, Italy

Unit of Occupational Health, Fondazione IRCCS San Gerardo dei Tintori, Monza, Italy
e-mail: michele.riva@unimib.it

L. Lorusso et al. (eds.), *Effects of Opera Music from Brain to Body*, Neurocultural Health and Wellbeing,
https://doi.org/10.1007/978-3-031-34769-6_8

Fig. 8.1 Giuseppe Permarini (1734–1808), portrait of the Austrian Martin Knoller (1725–1804), picture around half of the eighteenth century, Museo Teatro alla Scala, Milano

In the same period, Milan was one of the main centres of music and opera in Europe. The opera house *La Scala*, designed by architect Giuseppe Piermarini (1734–1808) and inaugurated in 1778, was as one of the leading opera and ballet theatres globally (Fig. 8.1).

In addition, the *Teatro Dal Verme*, inaugurated in 1872, was another regular venue for concerts and operas and also hosted some premières by Giacomo Puccini (1858–1924) and Ruggero Leoncavallo (1857–1919). Casa Ricordi, a publisher of primarily classical music and opera, was just founded in Milan in 1808 by violinist Giovanni Ricordi (1785–1853). It published the work of the major nineteenth-century Italian composers; its historical archives, at the Braidense National Library, contain music manuscripts and original letters by the major musicians of the last two centuries. Most of the Italian greatest opera composers had studied and worked in Milan: Giuseppe Verdi (1813–1901), Giacomo Puccini, Pietro Mascagni (1863–1945) and Ruggero Leoncavallo. Similarly, their librettists lived in that city, among which the best known was Francesco Maria Piave (1810–1876), who composed 10 librettos for Giuseppe Verdi, the most well-known being those for 'Rigoletto' and 'La Traviata'.

Hence, the composers and the librettists had certainly been influenced by the frenetic and industrial activities of the city, one of the world major financial and business centres and the industrial capital of Italy since the eighteenth century. So, it should come as no surprise that sounds and noises from the occupational environment can be found in the melodies and music of their compositions.

In this chapter, five tracks from Italian operas are selected and presented, testifying working conditions in the nineteenth and early twentieth centuries and disclosing some classical occupational diseases.

8.2 Sing when you Work!

'Il Trovatore' by Giuseppe Verdi.
Act 2, Scene 1.

Zingari	**Gypsies**
Vedi! Le fosche notturne spoglie	See! The endless sky casts off
De' cieli sveste l'immensa volta:	her somber nightly garb,
Sembra una vedova che alfin si toglie	just as a widow, who lays aside at last
I bruni panni ond'era involta.	the sad black veils of mourning.
All'opra! Dagli, martella!	So, to work! Strike, your hammers!
(danno piglio ai ferri del mestiere)	*(picking up their tools)*
Chi del gitano i sogni abbella?	Who brightens the gypsy's days?
La Zingarella!	The gypsy maiden!
Uomini	**Men**
(si rivolgono alle donne)	*(to the women)*
Versami un tratto:	Fill up the goblets:
lena e coraggio il corpo e l'anima	body and soul take strength
traggon dal bene	and courage from wine.
Tutti	**All**
Oh guarda, guarda!	Oh, look there, look there!
Del sole un raggio	A ray of sun
Brilla più vivido nel tuo bicchiere!	Flashes brighter in your glass
All'opra, all'opra…	So, to work! So, to work!
Chi del gitano i sogni abbella?	Who brightens the gypsy's days?
La Zingarella!	The gypsy maiden!
Zingari	**Gypsies**
Vedi! Le fosche notturne spoglie	See! The endless sky casts off
De' cieli sveste l'immensa volta:	her somber nightly garb,
Sembra una vedova che alfin si toglie	just as a widow, who lays aside at last
I bruni panni ond'era involta.	the sad black veils of mourning.
All'opra! Dagli, martella!	So, to work! Strike, your hammers!
(danno piglio ai ferri del mestiere)	*(picking up their tools)*
Chi del gitano i sogni abbella?	Who brightens the gypsy's days?
La Zingarella!	The gypsy maiden!
Uomini	**Men**
(si rivolgono alle donne)	*(to the women)*
Versami un tratto:	Fill up the goblets:
lena e coraggio il corpo e l'anima	body and soul take strength
traggon dal bene	and courage from wine.
Tutti	**All**
Oh guarda, guarda!	Oh, look there, look there!
Del sole un raggio	A ray of sun
Brilla più vivido nel tuo bicchiere!	Flashes brighter in your glass
All'opra, all'opra…	So, to work! So, to work!
Chi del gitano i sogni abbella?	Who brightens the gypsy's days?
La Zingarella!	The gypsy maiden!

'Il Trovatore' (The Troubadour) is an opera in four acts by Giuseppe Verdi to an Italian libretto by Salvadore Cammarano (1801–1852). It was first performed in Rome on 19th January 1853. The story takes place in Biscay and Aragon (Spain) during the fifteenth century and deals with the love rivalry between the gypsy Manrico (leader of the rebels against the king) and the Count di Luna, the commander of the royal forces.

The 'Coro di zingari' (Act 2, Scene 1), also known in English as the 'Anvil Chorus', depicts Spanish gypsies striking their anvils at dawn (hence its English name) and singing the praises of hard work, good wine and gypsy women.

The song of the gypsies leads us to make some comments on the relationship between music and work. Music has long been a uniting force among workers, since it can improve team spirit and provide an enjoyable diversion. According to Elbourne, in England during the early nineteenth century, the weavers were always 'singing away to the click of the shuttle' and even developed a rather saucy type of song [4]. In Italy, there has also been a long tradition of songs that 'mondine' (the rice-weeders) used to sing in the rice fields while working from dawn to sunset. Before the Industrial Revolution, labourers could work from home and pursue their musical hobbies too. In Northern Europe, the workers met in each other's homes or public houses to sing solo or in groups. The folk tradition within the workplace was replaced by the noise of the Industrial Revolution. When workers could no longer sing at work, they formed union ensembles and performed during their leisure time. So, many songs grew in response to trade union movement, especially in the twentieth century. Finally, during World War II, music at the workplace was aimed at keeping up morale and enhancing production [4].

But can music at the workplace increase work efficiency? In 1989, Mayfield and Moss conducted some studies to evaluate the effect of music tempo on task performance. Seventy undergraduate business students working for a stock market project were randomly divided into groups so that they listened to fast-paced music while they worked, to slow-paced music or to no music. Performance was significantly higher in the fast-paced music than in the slow-paced music, or no music [5].

Music exposure seems to have beneficial effects on training performance, even though it depends on different sound pressure levels, reaching better results at 70 dB SPL [6].

Shih et al. investigated the correlation between work concentration level and background music. In comparison with 'no music at all', those who had been listening to music prior to attention tests obtained higher scores in attentiveness (most probably a supplemental effect of music), whereas those who had been listening to music during the test showed a great variety of attention levels. So the findings seem to prove that background music positively affects people's job-site behaviour and workers' performances [7]. In case of complaints regarding music at workplace, workers and employers should consider the whole team feelings.

Even if some background music in the work environment can increase worker satisfaction and productivity, some studies indicate that noise may affect worker attention. In particular, music without lyrics is preferable because songs with lyrics are likely to reduce worker attention and performance [8].

In some contexts, music played may interfere with team communication. Music is played in 53–72% of surgical operations performed. Noise levels in the operating theatre already exceed World Health Organization recommendations. In these contexts, decisions around whether music is played and around the choice of music and its volume should be shared with team members [6].

Finally, it is worth recalling a legal case in Scotland, where the Performing Rights Society has recently brought a legal action against a firm, claiming that the company was infringing musical copyright by allowing employees to listen to music on the radio while at work. So, would any recorded music played at workplace fall under the same potential breach of copyright?

8.3 Noise and Confusion at the Forge

'Il Barbiere di Siviglia' by Gioacchino Rossi.
Act 1, Scene 16.

Tutti	**All**
Mi par d'esser con la testa	My head seems to be
in un'orrida fucina,	in an horrible forge:
dove cresce e mai non resta	where the importunate sound
dell'incudini sonore	of the anvils ceaseless
l'importuno strepitar.	and growing.
Alternando questo e quello,	Up and down, high and low striking,
pesantissimo martello,	the heavy hammer
fa con barbara armonia	with a barbarous harmony makes
mure e volte rimbombar.	the roofs and walls resound
E il cervello poverello,	Thus our poor, bewildered brain,
già stordito, sbalordito,	stunned, confounded,
non ragiona, si confonde,	without reason,
si riduce ad impazzar.	is reduced to insanity

'Il Barbiere di Siviglia' (The Barber of Seville) is an opera buffa in two acts by Gioachino Rossini (1792–1868) with a libretto by Cesare Sterbini (1784–1831). The première took place on 20th February 1816 in Rome. It was one of the earliest Italian operas to be performed in America and premiered in New York City in 1825.

During the seventeenth century in Sevilla (Spain), Count Almaviva asks Figaro, a clever barber, for assistance in helping him win and marry Rosina, ward of Dr. Bartolo who keeps the girl confined in his house [9].

At the end of Act 1 (Scene 16), Bartolo and his servants try and kick away Almaviva, disguised as a drunken soldier, as suggested by Figaro. Their hubbub has attracted a crowd and police arrive to silence the disturbers. When an officer is about to arrest him, Almaviva whispers his identity and is released to the astonishment of Rosina, Bartolo and his servants. To describe their state of confusion, they say their heads seem to be in a fiery smithy.

In this track, Rossini musically depicts the noisy work at the forge, the sound of the hammers and the astonishment of the minds, creating a sense of confusion also in the audience that is hard to keep up with the famous Rossinian crescendo of sounds and words. In addition to 'Il Barbiere di Siviglia', in another opera by Rossini, 'La Cenerentola' (Cinderella, 1817), referring to Cinderella's stepsisters, it is said that 'a forge is hammering the head of those women' (Act 1, Scene 1), and hence, the image of forge had already been used by the composer to indicate confusion and astonishment of the mind.

The forge is commonly considered as one of the noisiest workplace. The deafening sound of forging strokes by heavy machines threatens both the physical and mental health of many industrial workers [10]. In 1989 an analysis of a group of forge hammering workers demonstrated that they were exposed to impact noise ranging from 112 to 139 dB(A) – at an irregular rate of 20 to 50 drop/minute – and a continuous background noise ranging from 90 to 94 dB(A). Tinnitus and acoustic afterimage are related to the auditory effect of forge hammering noise [11]. In a study conducted in India in 2010, over 90% of workers engaged in various processes of casting and forging industry showed hearing loss in the noise-sensitive medium and higher frequencies. Occupation was significantly associated with NIHL, and hearing loss was particularly high among the workers of the forging section [12].

There is also a strong relationship between noise and mental confusion, reduced attentiveness and stress. Increased levels of stress are related to noise exposure during work. In particular, noise may also affect neuroendocrine activity, determining alterations in cortisol production [13]. Moreover, low-frequency noise might affect concentration and continuous and selective attention, especially in high-sensitive subjects [14]. Hence, Rossini's considerations on noise at the forge are not so far from reality.

Furthermore, recent studies have indicated that noise-induced stress could be the pathogenetic basis for other conditions, besides cognitive problems, already well depicted by Rossini. Indeed, in a wide spectrum of occupational factors, noise seems to be one of the major contributors to the increased risk of myocardial infarction [15]. In particular, the pathogenetic mechanism linking noise and cardiovascular diseases could be stress-related, but further investigations must be conducted to better evidence this relationship.

8.4 The Meaning of Work

'Cavalleria Rusticana' by Pietro Mascagni.

Act 1, Scene 4.

Alfio	**Alfio**
Il cavallo scalpita,	The horse stamps,
i sonagli squillano,	the bridle-bells jingle
schiocchi la frusta. Ehi là!	the whip cracks. Ho-La!

Alfio	**Alfio**
Soffi il vento gelido,	Let freezing winds blow
cada l'acqua e nevichi,	let it rain, let it snow,
a me che cosa fa?	what should I care?
Uomini	**Men**
O che bel mestiere	Oh what a fine trade
fare il carrettiere,	that of a carter,
andare di qua e di là!	travelling here and there!
Alfio	**Alfio**
M'aspetta a casa Lola	At home Lola awaits me,
che m'ama e mi consola	my love and my comfort
ch'è tutta fedeltà	who is all faithfulness
Donne	**Women**
Ah	Ah
Alfio	**Alfio**
Il cavallo scalpita,	The horse stamps,
i sonagli squillano,	the bridle-bells jingle
è Pasqua ed io sono qua!	it's Easter, and I am here.
Coro, Alfio	**Chorus, Alfio**
O che bel mestiere	Oh what a fine trade
fare il carrettiere,	that of a carter,
andare di qua e di là!	travelling here and there!

'Cavalleria Rusticana' (Rustic Chivalry) is a one-act opera by Pietro Mascagni to an Italian libretto adapted from a short story written by Giovanni Verga (1840–1922). Considered as one of the classic 'verismo' operas, it was first performed in Rome on 17th May 1890. This opera has figured in the sound track of several movies, most notably in the opening of 'Raging Bull' (1980) and in the conclusion of '*The Godfather Part III*' (1990).

In a Sicilian village, the young and charming Turiddu had returned from military service to find that while he was gone, his fiancée, Lola, had married Alfio, the prosperous village carter. However, Lola still loves Turiddu and has an affair with him. Alfio is made aware of her betrayal, so he challenges Turiddu to a duel and kills him.

In Easter morning (Act 1, Scene 4), Alfio, the carter, comes into the square, cheerfully singing the praises of his carter's life and also of the beauty of his wife Lola, still ignoring her betrayal. The praise of his own work, sung by Alfio, is an example of self-fulfilment and occupation positive value, highlighting the meaning of work in the past.

Since prehistoric era, men have drawn from their ability to work not only the livelihood but also the way to satisfy their own needs for creativeness, constructiveness and fulfilment. In ancient times, this need was very clear among craftsmen and was partially reduced by the advent of machinery during the Industrial Revolution [16].

What is now the meaning of work? Sapelli has written that work has a meaning only if the workers' psychophysical integrity is protected, and if by working, the individuals can get their own self-fulfilment through technical and professional competence. Without self-fulfilment, individuals may become depressed and lose the sense of living [17]. Pier Alberto Bertazzi (1945–2021), describing work as a 'basic human need and health promoting factor', has analysed some studies on the relationship between unemployment and lack of physical and mental health. The rise of the former, due to the recent world economic crisis, has led to an increase of anxiety disorders, depression and suicides [16]. Indeed, among the unemployed, the mortality increases for all causes and, particularly, for tumours and cardiovascular diseases, as shown also by a survey by Costa and Segnan in the Turin area [18]. Bertazzi has concluded that 'the role of occupational health physician is to recognize the possible negative effects of working condition and at the same time promote a positive approach to work, even in difficult conditions. [...] To be aware of the meaning of work makes work itself more liveable and more productive', contributing to the development of the economy and society [16].

Alfio's praises of his work and the beauty of his wife can well summarize the view of life as expressed by Sigmund Freud (1856–1939), who, to the question of what a healthy human should be able to do, answered by saying 'leiben und arbeiten' (loving and working) [19].

8.5 Health Problems among Gold Miners

'La Fanciulla del West' by Giacomo Puccini.

Act 1.

Minatori	**Miners**
Jim, perchè piangi?	Jim, why are you crying?
Jim! Che hai? Jim! Jim!	Jim! What's wrong? Jim! Jim!
Larkers	**Larkers**
Non reggo più, ragazzi!	I can't stand it any more, boys!
Mandatemi via! Son malato,	Send me away! I'm sick,
non so di che…Son rovinato!	don't know what it is. I'm finished!
Sono stanco di piccone e di miniera!	I'm fed up with pick-axe and mine!
Voglio l'aratro, voglio la madre mia!	I want the plough, I want my mother!
Sonora	**Sonora**
(raccoglie I soldi nel cappello)	*(taking a collection in his hat)*
Per rimandarlo a casa…	To send him home…
Minatori	**Miners**
Prendi…To'…Cinque dollari!	Here you are…take it…Five dollars!
Altri cinque!...A te, Son…	Another five! Take this, Son…
Anche questi…	This, too…
Sonora	**Sonora**
Coraggio!	Cheer up!
Larkens	**Larkens**
Grazie, grazie, ragazzi!...	Thank you, thank you, boys!...

Caricature of a *Fanciulla del West* rehearsal by Enrico Caruso

Fig. 8.2 Caricature of the *Fanciulla del West* (1910), rehearsal by Enrico Caruso (1873–1921). Metropolitan Opera Archives

'La Fanciulla del West' (The Girl of the West) is an opera in three acts by Giacomo Puccini to an Italian libretto based on a play by David Belasco (1853–1931). Following 'Madama Butterfly', it was first performed in New York City on tenth December 1910, under the direction of Arturo Toscanini (1867–1957) (Fig. 8.2).

The story takes place in a mining camp in the high Sierra Madre Mountains in California (USA), in 1849. Minnie, the girl who owns the saloon, courted by all the gold miners, falls in love with the bandit Dick Johnson who has been captured and must be hanged. Before his execution, Minnie convinces the miners that they owe her too much to let the man she loves be killed and so they set Johnson free.

In Act 1, suddenly, the miner Jim Larkens breaks down in the saloon, complaining of being unable to bear the harsh life in the gold mines. He is sick but does not know the nature of his disease. He would want to go back to work in the fields. One of the miners, Sonora, invites the others to give him money to go home.

What is the disease affecting Larkens? The miner's words show the tough life in Californian gold mines in the nineteenth and early twentieth centuries. In 2008, Ndlovu et al. reported autopsy findings in gold miners, died in Johannesburg between 1907 and 1913, just in the same years when Puccini was composing the opera. At that time there were high rates of death and disease in the mines mainly due to pneumonia, meningitis, tuberculosis and dysentery. Pneumonia and meningitis were the principle causes of death in new recruits and tuberculosis the main cause of mortality in referrals from the mines [20].

Another occupational disease was silicosis. In 1993, Hnizdo and Sluis-Cremer published data on 2235 South African gold miners who had, on average, 24 years of service from 1940 to the early 1970s, 14% of whom developed silicosis [21]. These data were not so different in the American gold mines [22]. Modern studies confirm that artisanal and small-scale miners, who work under archaic working conditions, are exposed to high levels of silica dust. Exposure to silica dust has been associated with an increased risk of tuberculosis and silicosis [23–25]. Obviously, in the case of Larkens, the latency period should be also considered, thus excluding some of these pathological conditions.

In addition to gold miners' health problems, when Puccini wrote 'La Fanciulla del West', he probably had in mind also the miners who constructed the Simplon tunnel (1906) linking Milan to Switzerland and France. Even if there had been only six fatalities from accidents, the workers of Simplon tunnel had suffered from diseases due to the rapid alternations of temperature, such as rheumatism, bronchitis and intestinal catarrh. However, it is well known that the precautions taken to ensure purity of drinking water had freed the miners from enteric fever and ancylostomiasis, thus avoiding the epidemic 'miners' anaemia' (or 'miners' cachexia'), which instead had occurred during the construction of Saint Gotthard tunnel (1883) [26].

Finally, the disease of Larkens can be more simply interpreted as homesickness or '*nostalgia*'. This condition was looked upon as an occupational disease when it was first described by the Alsatian Johannes Hofer (1669–1752) in his 'Dissertatio Medica de Nostalgia' (1688), defining it as 'the sadness engendered by the passionate desire to return home' among the Swiss mercenaries who in the plains of lowlands of France or Italy were pining for their native mountain landscapes. Indeed, it was also known as 'mal du Suisse' (Swiss illness) or 'Schweizerheimweh' (Swiss homesickness). Soldiers were sometimes successfully treated by being dismissed and sent home. From the 1850s, nostalgia was no longer considered as a specific disease but rather as a symptom or stage of a pathological process.

Loneliness is another factor that could be connected with psychological distress in workers. In 2016, a study on Chinese rural-to-urban migrant workers who were employed in service industry evidenced that they were generally more marginalized in cities. Social isolation caused an increase of the feelings of loneliness [27]. The condition of these migrant workers does not seem so different from that of Larkens and the other gold miners represented by Puccini. Finally, the current coronavirus (COVID-19) pandemic has had large impacts on society, including people practicing social distancing [28]. The effects of loneliness that affects occupational mental health are still being studied.

8.6 Psychological Problems on the Stage

'Pagliacci' by Ruggero Leoncavallo.
Act 1, Scene 4.

Canio	**Canio**
Recitar!	Perform the play!
Mentre preso dal delirio,	While I'm out of my mind,
non so più quell che dico	not knowing what I say
e quel che faccio!	or what I do
Eppur…è d'uopo…sforzati!	And yet. I must…ah, force myself to do it!
Bah, sei tu forse un uom!	Bah! You are not a man!
Tu sei Pagliaccio!	You are Pagliaccio!
Vesti la giubba e la faccia infarina:	Put on the costume and the powder:
la gente paga e rider vuole qua.	the people pay and want to laugh.
E se Arlecchino t'invola Colombina,	And if Arlecchino steals your Colombina,
ridi Pagliaccio, e ognun applaudirà!	laugh Pagliaccio, and all will applaud you!
Tramuta in lazzi lo spasmo e il pianto;	Change tears and anguish into clowning;
in una smorfia il singhiozzo e il dolore…	and into a grimace your sobbing and pain…
Ridi, Pagliaccio, sul tuo amore infranto!	Laugh, Pagliaccio, at your shattered love!
Ridi del duol che t'avvelena il cor!	Laugh at the sorrow that has rent your

'Pagliacci' (Clowns) is an opera in two acts and a prologue written and composed by Ruggero Leoncavallo (1857–1919). It premiered at the Teatro Dal Verme in Milan on 21st May 1892, being conducted by Arturo Toscanini (Fig. 8.3).

Fig. 8.3 Portrait of Enrico Caruso as Canio in *Pagliacci* from a postcard published circa 1904

The story deals with the tragedy of a jealous husband in a theatrical troupe in Calabria (Italy), during the Feast of the Assumption, between 1865 and 1870. Canio, head of the troupe (playing the role of Pagliaccio in Commedia) is aware that his wife Nedda is in love with Silvio; like in the play Colombina, Pagliaccio's wife, loves Arlecchino. On the stage, Canio costumed as Pagliaccio, forgetting the play, grabs a knife and stabs to death Nedda, costumed as Colombina, and his lover, Silvio.

In Act 1, Scene 4, before performing a comic play, when changing into costume, Canio, aware of the betrayal of his wife, feels unable to perform while suffering. He steels himself to face the public with a smile even though his heart is breaking: the clown must laugh.

The clown is no longer able to entertain people, due to his psychological problems, worries and suspicions. Leoncavallo's Canio recalls the figure of Rigoletto in the homonymous opera by Giuseppe Verdi (1851).

In Act 1, Scene 4, Rigoletto, the hunchbacked court jester, says 'Oh mankind! Oh nature! It was you who made me evil and corrupt! I rage at my monstrous form, my cap and bells! To be permitted nothing but to laugh! I'm denied that common human right, to weep. My master, young, carefree, so powerful, so handsome, half-dozing, says:—Fool, make me laugh!—And I must contrive to do it! Oh, damnation!'

As well as Rigoletto, Canio, a comedian, must make people laugh, even if he does not feel like and admits he cannot do it. It is a clear example of emotional dissonance that may predict emotional exhaustion and depersonalization.

Depression and exhaustion of emotional strength and motivation are usually the result of prolonged stress or frustration. The killing spree and delirium unleashed on stage by the jealousy of Canio, confounding play and real life, is an extreme case, but there are many other psychological disorders affecting actors, musicians and singers on stage. Performance anxiety, or stage fright, was defined as 'anxiety aroused by potential mishaps in performance that expose feared inadequacies before an audience and which evoke feelings of embarrassment and humiliation' [29]. For affected musicians, performance anxiety can be emotionally devastating, as their career choice in music may be terminated or severely compromised and this is especially evident in opera performers. In a recent study, the relationship among trait anxiety, occupational stress, perfectionism, aspiration and music performance anxiety was investigated in a group of elite operatic chorus artists employed full time by a national opera company. The chorus artists reported higher trait anxiety, higher occupational role concerns and higher occupational personal strain than normative samples [30].

References

1. Riva MA, Sironi VA, Tremolizzo L, Lombardi C, Devito G, Ferrarese C, Cesana G. Sleepwalking in Italian operas: a window on popular and scientific knowledge on sleep disorders in the 19th century. Eur Neurol. 2010;63:116–21.
2. Fustinoni S, Riva MA. The first steps of industrial hygiene and occupational toxicology at the Clinica del Lavoro di Milano under the guidance of Luigi Devoto. Med Lav. 2019;110(S1):24–35.

3. Riva MA, Belingheri M, Fustinoni S. The contribution of the Clinica del Lavoro of Milan to the development of industrial hygiene and toxicology in the twentieth century. Arch Environ Occup Health. 2019;74:30–41.
4. le Roux GM. "whistle while you work": a historical account of some associations among music, work, and health. Am J Public Health. 2005;95:1106–9.
5. Mayfield C, Moss S. Effect of music tempo on task performance. Psychol Rep. 1989;65:1283–90.
6. Nees LK, Grozinger P, Orthmann N, Deutsch TM, Hennigs A, Domschke C, Wallwiener M, Rom J, Riedel F. Evaluating the influence of music at different sound pressure levels on medical students' performance of standardized laparoscopic box training exercises. BMC Med Educ. 2021;21(1):209.
7. Shih YN, Huang RH, Chiang HS. Correlation between work concentration level and background music: a pilot study. Work. 2009;33:329–33.
8. Shih YN, Huang RH, Chiang HY. Background music: effects on attention performance. Work. 2012;42(4):573–8.
9. Riva MA, Bellani I, Turato M, Cesana G. Physicians and alternative medicines in "the barber of Seville" by Gioachino Rossini: a bicentennial debate. Eur J Intern Med. 2015;26(10):757–8. https://doi.org/10.1016/j.ejim.2015.09.018.
10. Razavi H, Ramazanifar E, Bagherzadeh J. Optimizing noise control strategy in a forging workshop. Int J Occup Saf Ergon. 2014;20(2):257–64.
11. Kamal AA, Mikael RA, Faris R. Follow-up of hearing thresholds among forge hammering workers. Am J Ind Med. 1989;16:645–58.
12. Singh LP, Bhardwaj A, Kumar DK. Prevalence of permanent hearing threshold shift among workers of Indian iron and steel small and medium enterprises: a study. Noise Health. 2012;14:119–28.
13. Waye KP, Bengtsson J, Rylander R, Hucklebridge F, Evans P. Clow a low frequency noise enhances cortisol among noise sensitive subjects during work performance. Life Sci. 2002;70:745–58.
14. Pawlaczyk-Luszczyńiska M, Dudarewicz A, Waszkowska M, Szymczak W, Sliwińska-Kowalska M. The impact of low-frequency noise on human mental performance. Int J Occup Med Environ Health. 2025;18:185–98.
15. Bortkiewicz A, Gadzicka E, Siedlecka J, Szyjkowska A, Viebig P, Wranicz JK, Kurpesa M, Dziuba M, Trzos E, Makowiec-Dąbrowska T. Work-related risk factors of myocardial infarction. Int J Occup Med Environ Health. 2010;22:1–11.
16. Bertazzi PA. Il lavoro come bisogno umano e fattore di salute [work as a basic human need and health promoting factor]. Med Lav. 2010;101(Suppl 2):28–43.
17. Sapelli G. Il lavoro ha un significato? Fondamenti per una cultura positiva del lavoro [does work have a meaning? Basis for a positive work culture]. Med Lav. 2010;101(Suppl 2):7–10.
18. Costa G, Segnan N. Mortalità e condizione professionale nello studio longitudinale torinese [Mortality and occupational status in the Turinese longitudinal study]. Epidemiol Prev. 1988;10:48–57.
19. Bertoli M. Il lavoro come spazio per una autostima possibile [Work as a possible means of self-esteem]. Med Lav. 2010;101(Suppl 2):78–81.
20. Ndlovu N, Murray J, Davies A. Autopsy findings in Witwatersrand gold miners, 1907-1913. Adler Mus Bull. 2008;34:3–12.
21. Hnizdo E, Sluis-Cremer G. Risk of silicosis in a cohort of White south African gold miners. Am J Ind Med. 1993;24:447–57.
22. Steenland K, Brown D. Silicosis among gold miners: exposure-response analyses and risk assessment. Am J Public Health. 1995;85:1372–7.
23. Moyo D, Zishiri C, Ncube R, Madziva G, Sandy C, Mhene R, Siziba N, Kavenga F, Moyo F, Muzvidziwa O, Ncube P, Chigaraza B, Nyambo A, Timire C. Tuberculosis and silicosis burden in artisanal and small-scale gold miners in a large occupational health outreach Programme in Zimbabwe. Int J Environ Res Public Health. 2021;18(21):11031.

24. Knight D, Ehrlich R, Cois A, Fielding K, Grant AD, Churchyard G. Predictors of silicosis and variation in prevalence across mines among employed gold miners in South Africa. BMC Public Health. 2020;20(1):829.
25. teWaternaude JM, Ehrlich RI, Churchyard GJ, Pemba L, Dekker K, Vermeis M, White NW, Thompson ML, Myers JE. Tuberculosis and silica exposure in south African gold miners. Occup Environ Med. 2006;63(3):187–92.
26. Hobson J. ICOH one hundred and the Simplon tunnel. Occup Med. 2006;56:221–2.
27. Zhong B, Xu Y, Jin D, Zou X, Liu T. Prevalence and correlates of loneliness among Chinese service industry migrant workers: a cross-sectional survey. Medicine (Baltimore). 2016;95(24):e3903.
28. Fujii R, Konno Y, Tateishi S, Hino A, Tsuji M, Ikegami K, Nagata M, Yoshimura R, Matsuda S, Fujino Y. Association between time spent with family and loneliness among Japanese workers during the COVID-19 pandemic: a cross-sectional study. Front Psych. 2021;12:786400.
29. Nagel JJ. Treatment of music performance anxiety via psychological approaches: a review of selected CBT and psychodynamic literature. Med Probl Perform Art. 2010;25:141–8.
30. Kenny DT, Davis P, Oates J. Music performance anxiety and occupational stress amongst opera chorus artists and their relationship with state and trait anxiety and perfectionism. J Anxiety Disord. 2004;18:757–77.

Playing and Singing: A Demanding Profession

9

Michael Belingheri, Maria Emilia Paladino, and Michele A. Riva

9.1 Historical Notes

In Ancient Greece, music was a crucial factor for the good education of a citizen; the combination of song and dance was considered the basics for harmony. Vocal practice played a decisive role in the education of good citizens, who were often chosen as members of choir during festivals in the cities. During their training, chorists often used herbs and potions to improve their voice. In the speech "On the Choreutes," the Attic orator Antiphon of Rhamnus (480–411 BC) referred to the death of a boy, Diodotus, who was being trained to sing in a choir and was accidentally poisoned by a drug given him to improve his voice. The choirmaster was accused of poisoning before the Areopagus, the court that tried cases of deliberate homicide in Athens. Ancient physicians also paid attention to the care of the voice of singers. For example, in his treatise "On Diet," Hippocrates (c. 460–370 BC) recommended to avoid some exercises that could harm the voice. Even if they were not described, musculoskeletal problems of musicians existed even in antiquity: the string instruments of the Greek musicians were fairly heavy, while the musical and dramatic performances lasted several hours.

In Rome, during the first century AD, Scribonius Largus indicated that licorice was a valid remedy for problems of the voice. In medieval monasteries, monks singing Gregorian chants used herbs and other natural remedies to heal their voices. In the Early Modern Age, music began to branch out beyond the church music that had dominated before this time. Opera arose at this time in Italy and professional singers began to appear on theatre stages in all over Europe. Their voice disorders and other

M. Belingheri · M. E. Paladino · M. A. Riva (✉)
School of Medicine and Surgery, University of Milano-Bicocca, Monza, Italy

Unit of Occupational Health, Fondazione IRCCS San Gerardo dei Tintori, Monza, Italy
e-mail: michele.riva@unimib.it

L. Lorusso et al. (eds.), *Effects of Opera Music from Brain to Body*, Neurocultural Health and Wellbeing,
https://doi.org/10.1007/978-3-031-34769-6_9

specific diseases attracted the attention of physicians and scholars of the time. In Padua, the renowned anatomist Gabriele Falloppio (1523–1562) stated: "Singers with deep voices, commonly called bass [...] usually suffer from hernias because they sing without interruption and have to use their abdominal muscles to produce this resonance and sing loudly" [1]. In the same period, Girolamo Mercuriale (1530–1606) observed that "pitching the voice high leads to swelling in the head, throbbing temples, pulsations of the brain, swelling of the eyes and a tinnitus of the ears" (De Arte Gymnastica," 1569) [1]. At the beginning of the eighteenth century, Bernardino Ramazzini (1633–1714), founder of the Occupational Health, dealt with diseases of workers, including singers. In chapter XXXVIII of his masterpiece "De Morbis Artificum Diatriba" (1700), Ramazzini described the diseases that generally afflict music teachers and singers (Fig. 9.1).

In particular, he sustained that head colds and hoarseness were common ills in singers, once again owing to the muscle tension that caused more lymph than necessary to be secreted by the salivary glands [1]. Referring to Margherita Salicola (1660–1717), a renowned opera singer of that time, he wrote that "after singing her part, she very often has a severe attack of hoarseness" and added that "after she sang high notes on the stage for some times [...] she usually has an attack of vertigo soon after." Ramazzini also stated that singers usually suffer from hernias, because when singing, they have to breathe deeply and hold in the air for a long time, so that the abdomen muscles that aid respiration relax, thus facilitating inguinal hernias. Ramazzini concluded his analysis with the category of flautists and any other players of wind instruments. Owing to their considerable exertions while blowing their trumpets and flutes, they are not only afflicted by the aforementioned ills, but also by much more serious ones, such as rupture of pulmonary vessels and sudden discharge of blood from the mouth [1].

Fig. 9.1 Bernardino Ramazzini (1633–1714), Italian physician and founder of Occupational Medicine (portrait of 1717) and his Frontpage of the definitive edition "*De Morbis Artificum Diatriba*" (Padua, Conzatti, 1713)

Ramazzini also discussed the health problems of *castrati*. From the end of the sixteenth to the whole eighteenth century, the prepuberal castration was carried out to preserve the young male voice into adult life. Gaspare Pacchierotti (1740–1821) was a celebrated Italian castrati of the eighteenth century and one of the most famous opera singers in the history. A recent analysis of his human remains showed that Pacchierotti's cervical vertebrae were all strongly eroded [2]. In singing, a correct postural attitude requires the cervical spine to be maintained in an erect position, in nuchal elongation, avoiding the lordosis, the extension and lifting of the jaw. This prolonged position in Pacchierotti, combined with the osteoporosis related to hormonal alterations in castrati, can account for the eroded states of his cervical vertebrae. This study was one of the first researches to evidence the occupational markers in remains of a professional singer [2] (Fig. 9.2).

During the nineteenth century, tuberculosis was a typical disease of opera singers and, generally, of actors on stages. Singing has been noted as a possible means of disseminating tuberculosis; the droplets produced during this activity are small and therefore occasionally dangerous [3]. According to Linda and Michael Hutcheon, "tuberculosis is perhaps the perfect operatic disease: it involves the breath—as both inspiration and expiration, as both site of song and locus of disease" [4]. Likely, opera singers identified themselves with consumptive heroines, such as Violetta from "*La Traviata*" by Verdi and Mimì from "*La Bohème*" by Puccini. They were frightened of a disease affecting their working tool, the lungs, and well knew its symptoms and physical signs. So on the stage they could best interpret the "*spes phthisica*" (consumptive euphoria) preceding Violetta's death or the "*gelida manina*" (the frozen little hand) and the persistent cough of Mimì. Similarly, singers

Fig. 9.2 The castrato Gaspare Pacchierotti (1740–1821), a portrait of Pacchierotti in his fifties of an unknown painter around 1790–1800

were scared of the tragic end of the phthisical Antonia, who, in "*The Tales of Hoffmann*" by Offenbach, was induced by the ghost of her mother to sing, with fatal results [4].

Until the beginnings of the twentieth century, doctors mainly continued to study only voice disorders in singers, as shown by a paper on this issue published in 1867 by Malachia De Cristoforis (1832–1915), one of the founders of the modern Occupational Health in that period in Milan [5]. The only exemption was the study by the British physician George Vivian Poore (1843–1904) on musician's cramp. He made the "Clinical lecture on certain conditions of the hand and arm which interfere with the performance of professional acts, especially piano-playing" at the Royal College of Surgeons [6]. Analyzing etiopathogenesis, clinical manifestations, and experience of treating 21 professional musicians, he concluded that "the most important point in treatment is rest." In the same years, Adolphe Wahltuch (1803–1897), another British physician, used electrotherapy to treat musculoskeletal problems in young female violinists with a certain degree of success [7].

In the first decades of the twentieth century, physicians started to be interested in disease of all musical performers, not only singers. In particular, in his work "*Die Berufskrankheiten der Musiker*" (1926), the German neurologist and musicologist Kurt Singer (1885–1944) first systematically described symptoms of musicians' diseases and their treatment [8] (Fig. 9.3).

In the preface of his work, Singer stated: "this book, the first of its kind ever to be published, treats of the diseases, *both mental and physical*, that are peculiar in the musical profession" [8]. Singer's treatise was particularly focused on mental problems of musicians, probably influenced by new Freudian theories. He identified a connection between musicians and hysteria: "the musical work itself creates a terrible strain on the apparatuses of thought and fantasy, linked to a heightened deterioration of the nervous substance" [8].

Fig. 9.3 The German Neurologist, musicologist, and conductor Kurt Singer (1885–1944)

In the second part of the twentieth century, the health problems of musicians have become more formally recognized by interested clinicians. A new interest began in the 1980s with the publication of several pioneering surveys of orchestral musicians, which revealed the high prevalence of these disorders and stimulated research in this field [9]. The great opera houses started to worry about the health problems of singers, musicians, and choir members, providing them medical visits and access to healthcare facilities. Occupational Health physicians, specialists in audiology and phoniatrics, physiatrists, and orthopedic surgeons started to study opera performers and their diseases.

In this chapter, the main disorders of professional performers are analyzed, including voice and hearing disorders and musculoskeletal and psychological problems. Finally, the impact of the recent COVID-19 pandemic on the internal organization of opera houses and the health of their employees is discussed.

9.2 Voice Disorders

Professional singers depend on their voices to pursue their profession. They need healthy vocal tracts with great endurance to manage the great vocal load at work. As demonstrated by the aforementioned historical examples, professional singers and other vocal performers in opera houses may suffer from functional vocal disorders. The most common of these disorders are vocal fatigue and muscle tension dysphonia or aphonia. The vocal fatigue is caused by the overuse of the voice and resultant tiring of the laryngeal musculature. The muscle tension dysphonia or aphonia is caused by hypertonicity of the laryngeal musculature, which limits the vocal folds' ability to abduct and adduct with coordination and rapidity.

In classical choir singers or professional opera choristers, self-reported voice problems were detected in 38–43.6% [10, 11]. In another study, the prevalence of vocal fatigue in a 10-year period among 250 opera singers examined was 22% [12]. The three most frequent symptoms observed in the study group were muscle pain (87.27%), muscle fatigue (76.36%), and diffuse sharp pain (70.37%). The most frequent complaints were pain-related, especially in the female singer group. The most common vocal complaints were whispery voice (63.64%) and hoarseness (61.82%). More than a half of opera singers (52.73%) had difficulties with maintaining the vocal pitch during singing and a quarter (21.82%) noticed changes in vibrato. Among risk factors, we have found excessive effort during performance; inadequate vocal rest regime; insufficient training; inappropriate singing warm-up; high-pitched, forced phonation in a short period; and prolonged singing beyond the appropriate tessitura (most acceptable and comfortable vocal range for the given singer) [12]. Opera singers seem to report heartburn, regurgitation, coughing, and hoarseness more often than the controls [13].

Table 9.1 shows the prevalence of self-perceived voice problems in 257 vocal performers [14].

Tessitura seems to be a significant risk factor: the higher voices are at a heightened risk of having voice disorders. In a study that examined 70 opera singers (30

Table 9.1 Frequency of symptoms of vocal impairment ($N = 257$)

Symptom of vocal impairment	Frequency	% of subjects
Hoarseness	193	75.1
Voice fatigue	178	69.3
Tickling/choking sensation while speaking	138	53.6
Pain in throat	135	52.5
Change in pitch of speaking voice	114	44.3
Difficulty with volume/projection	90	35.0
Lump in throat feeling while speaking	73	28.4
Strained or strangled voice	73	28.4
Voice breaks or stops when talking	71	27.6
Complete loss of voice	30	11.6

females, 40 males; aged 19–70 years), the prevalence of voice disorders was the highest among sopranos, i.e., 50%, in comparison to 20% among countertenors, tenors, and baritones, 10% among contraltos and mezzo-sopranos, and no cases among basses [15].

Organic laryngeal pathologies include phonotraumatic benign lesions like nodules ("singer's nodes"), polyps, pseudocysts, cysts, hemorrhage, vascular ectasia, sulcus vocalis, hematomas, and vocal edema. Such lesions related predominate in professional performers rather than nonprofessional singers. Actually, the technical abilities and laryngeal architecture of voice professionals might mitigate the risk of developing these conditions [16].

It is important for the population of opera singers to evaluate the actual risk that they are exposed to over the years of their career, which might encourage them to control their vocal folds more often than other professional categories and comfort their vocal folds with regular voice rehabilitation.

In addition to vocal problems, some chemical substances on stages could harm the upper and lower respiratory tract. For example, cobalt and aluminum could be pigment components in a freshly painted stage set. Pigment particles, rubbed off because of frequent set changes, may contaminate stage dust and thus indirectly expose the singers to the harmful substances. Diisocyanates (DII) are contained in decorative foams frequently used as materials for stage sets, being malleable. All these substances may irritate or damage respiratory tracts [17]. In addition, in opera houses, singers currently complain about a dry, dusty environment, likely to expose them to respiratory tract problems.

9.3 Hearing Disorders

Over the past 50 years, hearing problems in professional musicians have been studied. In particular, music-induced hearing loss was seen in up to 58% of classical musicians. In addition, permanent or temporary shifts of hearing thresholds, hyperacusis, loss of hearing sensitivity, and tinnitus were observed [18].

Hearing loss seems to be provoked by both extrinsic and intrinsic factors [19, 20]. Extrinsic factors could be represented by number of years of exposure to high levels of sound. The number of hours of music exposure per week has been found to have greater effect in predicting hearing loss in both ears. Other extrinsic factors are playing in acoustically unsuitable venues, the position of the players in the orchestra on stage, and in an orchestra pit [18]. For example, musicians located near percussion instruments (e.g., drums) and brass instruments (e.g., trumpet, trombone) tend to have higher exposures and more hearing disorders [18]. Moreover, the instruments being played and their sound levels are another factor that should be considered (Table 9.2) [21]. Intrinsic factors related to individual differences in psychophysiological proneness to hearing disorders among musicians appear less studied [18].

Tinnitus is often one of the first hearing problems in musicians. A higher risk of hearing loss was observed in musicians who had experienced acute symptoms of exposure to loud noise, such as tinnitus or ear pain [22]. In a study conducted among musicians with a minimum of 5 years of experience, 50% of participants stated that they had experienced tinnitus after a performance and 28% had ear pain during the performance; 56% of the subjects had experienced one of these symptoms during or after the performance [23]. Studies have shown that 39% of all the musicians suffer from tinnitus, mainly localized to the left ear. The duration of tinnitus is another factor that should be considered: 24% of professional orchestra musicians reported of tinnitus for at least 5 min in duration and 50% of them reported experiencing constant tinnitus [24]. Other studies have reported tinnitus in 39% and 45% of musicians [25].

Other common audiological symptoms in musicians are hyperacusis and diplacusis. Hyperacusis, defined as a reduced tolerance to sounds of average intensity, sometimes accompanied by painful sensitivity to ordinary environmental sounds, was found in 20% of musicians in different studies [26]. Diplacusis, an anomaly whereby the same tone is perceived as having a different pitch depending on whether

Table 9.2 Sound levels of different instruments

Sound levels of music	
Instrument	**dB**
Normal piano practice	60–70
Piano fortissimo	84–103
Violin	82–92
Cello	85–111
Oboe	95–112
Flute	92–103
Piccolo	90–106
Clarinet	85–114
French horn	90–106
Trombone	85–114
Timpani and bass drum	106

it is presented in the right or in the left ear of the same listener, is rarer and was evidenced in 6% of investigated musicians [26].

Earing protection devices (EPDs), such as foam plugs, could help to mitigate the effects of noise on hearing, but they are scarcely accepted by musicians. In a study by Dinakaran et al., printed in 2018, 70% of musicians were not aware of EPDs and the remaining participants never used them as they did not know where to purchase the EPDs from and their importance in hearing conservation [22]. Only 22% of college music students reported wearing EPDs when exposed to potentially harmful sound levels. Some musicians seem to use occasionally hearing protection and in one ear only [22].

So, it is important to educate orchestra musicians regarding the effects of exposure to loud music on the hearing system and to change their attitudes towards the use of hearing protectors.

9.4 Performance-Related Musculoskeletal Disorders

The acronym "PRMD" (performance-related musculoskeletal disorders) is used to defined "pain, weakness, numbness, tingling or other symptoms that interfere with the ability to play the instrument at the level one is accustomed to" [27]. A systematic review, conducted in 2015, on different studies among musicians concluded that lifetime prevalence of pain affecting the playing capacity varied between 25.8% and 84.4% [27].

Common PRMDs of musicians include overuse problems, such as tendonitis and peripheral nerve entrapment syndromes, that typically affect the upper extremities, the neck, the back, and the facial musculature. In particular, in the aforementioned review of 2015, the most affected body regions were the low back (9.8% to 66.7%) and neck (9.8% to 48.5%). These conditions often become chronic, painful, and disabling health problems that last for years [28]. A recent meta-analysis tried to determine the prevalence of temporomandibular disorders (TMDs) in musicians, estimating a prevalence of 52.8% for wind instruments and 53.9% for string instruments [29].

In an orchestra, string players experience more musculoskeletal disorders than other instrumentalists. Indeed, the odds of wrist/hand pain were 2.9-fold higher in musicians who play string instruments, than for those who play wind instruments, while the odds of elbow pain were 50% lower among wind and brass players. Several studies have found female musicians to have a higher rate of injury than males [30–32]. No conclusions can be made regarding the association of age, number of years playing, and the possible protective role of physical or musical warm-up. Among risk factors, the average number of hours played per week was evidenced by different studies; performance anxiety and work-related stress seemed to be positively related with musculoskeletal disorders in musicians. In particular, some psychosocial aspects of work, such as long hours at work, work content, high job demands, low control/influence, and a lack of social support seems to be related to

Table 9.3 Common PRMD related to specific instrument

Instrument	Common musculoskeletal disorders
Piano	Muscle strain in hands, extrinsic flexors, and extrinsic extensors, ulnar wrist pain, carpal canal syndrome, shoulder weakness, and scapular instability
Woodwinds and brass	Wrist and thumb pain, De Quervain's tenosynovitis, bilateral thoracic outlet-like symptoms, carpal tunnel syndrome
Percussion	Tenosynovitis and arthritis in bilateral hands and wrists, carpal tunnel syndrome, intervertebral disc problems, muscle spasms
High-string instruments	Neck and shoulder pain, scapular dyskinesia, strain on the forearm, finger, and thumb muscles
Low-string instruments	Thumb, radial wrist, and forearm pain, digital nerve irritation, and paresthesia at the tip of the finger

Table 9.4 The four components of MPA

Affect	Feelings of anxiety, tension, apprehension, dread, or panic
Cognition	Loss of concentration, heightened distractibility, memory failure, catastrophizing
Behavior	Tremor, difficulty in maintaining posture and moving naturally
Physiology	Disturbances in breathing pattern, dry mouth, cardiovascular changes, gastrointestinal disturbances

PRMD [33]. Similarly, public exposure, competition, job context, and criticism could be important stressors that affect classical instrumental musicians [34].

Another important risk factor is the type of instrument, as shown in Table 9.3 [35].

As shown for voice and hearing disorders, prevention and treatment are essential in these kinds of health problems. To prevent these disorders, the effect of posture has been explored in some musicians, such as saxophone players [36]. Education in proper posture is important to avoid head/neck rotation and back pain. No clear results were found in studies that analyzed the effects of exercise on PRMD prevalence. For example, in some researches increased exercise appears to be associated with less risk of PRMDs [37], while other studies did not evidence this effect [38, 39]. Physical therapy and occupational therapy seems to be useful in the treatment of these problems.

9.5 Music Performance Anxiety

In the Diagnostic and Statistical Manual of Mental Disorders (DSM-5), music performance anxiety (MPA) is classified as a subtype of social anxiety disorder. It is one of the most frequently reported disorders among professional performers, with an estimated prevalence rate between 15% and 25% [40]. Performers generally suffer from this problem before and during performances, while about 21% suffer from anticipatory anxiety [41].

The four components of this disorder are shown in Table 9.4 [42]; it occurs on a continuum of severity from normal stress and anxiety intrinsic to the profession to debilitating symptoms of "stage fright" and panic.

A recent review, published on 2018, summarized scientific literature on risk factors and treatment effects of MPA among professional musicians [43]. Regarding risk factors, women seem to be more affected than men, like what happened with other anxiety disorders. Older musicians appear to be less affected by this disorder, since younger performers are more vulnerable, also due to problems related to career entry or financial uncertainties. On the contrary, MPA seems to be rather independent of musical training and expertise. No differences between type of instrument, type of orchestra, or the position within the instrument group were found. As a general factor, anxiety as an innate personality trait of the individual contributes to the development of this problem. MPA increased with the size of the audience and the difficulty of the performance [43].

Concerning treatment, cognitive behavioral therapy appears to provide beneficial results, evidencing reductions of MPA after interventions. An effective pharmacological treatment option is β-blockers. β-Blockers reduced the vegetative symptoms of MPA, such as heart rate and tremor, but they were not effective on anxiety, negative cognitions, and behavior. More research is needed to clarify the effects of other interventions, such as music therapy improvization and desensitization, yoga and relaxation, psychodynamic therapy, and hypnotherapy [43].

9.6 Music and COVID-19: Reinventing Opera

In December 2019, a novel coronavirus was identified in Wuhan (China) and designated as SARS-CoV-2. It caused some important clinical manifestations that were named coronavirus disease 2019 (COVID-19) [44–46]. The World Health Organization (WHO) declared a pandemic on March 11, 2020, so social life was shut down in several countries and most theatres and opera houses were closed. La Scala in Milan has been closed since February 23, 2020, while the Metropolitan Opera in New York since March 12, 2020.

When the restrictions have been cancelled, all the major stage productions started to be scheduled, mainly in the 2020/2021 season. With the opening of theatres, the main aim was to ensure performances and general artistic work at the opera house without increasing the risk for the artistic staff and the audience being infected. The main guidelines applied in theaters and opera houses worldwide are reported below.

In particular, the guidelines to audience were personalized ticket sales, separation of audience and performers, regulated entrance and exit, the obligation to wear a surgical face mask, and a limited number of people in the theatre to guarantee a distance of 1.5 meter between the auditors. Actually, due to the increasing incidence of cases in the last months of 2020, performances generally occurred without audiences and were only transmitted virtually [47].

Different guidelines were applied to performers, who continued to sing and play instruments in rehearsals and on stage. One of the problems was artistic staff could not comply with protective measures because of their instrumental, singing, and dancing activities. They could not maintain a distance of 1.5 m and wear protective masks. Furthermore, singers and wind instruments performed aerosol-forming activities. For high-risk groups of performers, appropriate measures were introduced: mouth-nose protection on stage, intensive ventilation, and cleaning. In the orchestra pit, musicians were positioned at the distance of 1.5 m to the next player and 2 m in the direction of the wind for wind players. If the orchestra pit did not allow to maintain distance, rows of stalls were removed and the musicians played in an expanded space [47].

In the choir, the minimum distance of 2 m in the direction of singing was required. The music sheets were given individually and were not exchanged among each other. It should be noted that high COVID-19 transmission rates in choirs with deaths in closed rooms have been reported [48]. Good ventilation conditions in the rehearsal rooms and on the stage could prevent the transmission among singers.

Choreographies were also modified for limited groups of dancers. Indeed, there is a high risk of infection among the dancers, probably due to physical proximity and deep exhalation during physical exercise, which produces increased level of aerosols [47].

Finally, makeup and hairdressing services were to be as brief as possible and performed in strict compliance with all hygiene regulations.

When maintaining the distance of 1.5 m between the performers was not feasible, staff members were tested consecutively using PCR swabs. The testing frequency was adapted continuously depending on epidemiological contest and artistic requirements. Furthermore, in case of symptoms or close contact with a COVID-19-positive person in private or professional environment, the staff members were tested. In case of outbreaks, immediate quarantine of the risk group prevented further infections among artists [47].

In the second part of 2021, the vaccination against COVID-19 modified restrictions and guidelines for vaccinated people. In Italy, a digital COVID Certificate (EU Green Pass) is required to access theatres from August 2021. Similarly, the proof of vaccination is required to access other important opera houses, such as the Metropolitan Opera House in New York, Sydney Opera House, and Royal Opera House in London. In compliance with the CDC guidelines for fully vaccinated individuals, the audience is at full capacity.

Finally, the impact of COVID-19 on infected musicians and singers should be mentioned. Indeed, COVID-19 can lead to dysphonia, caused by a damaged vocal fold function [49]. Furthermore, the infection could cause permanent lung dysfunction with a reduction in pulmonary function, which could be career-ending for a professional singer or a wind instrument player [49]. Vaccination of professional performers could be useful to prevent the infection and its severe consequences [50].

References

1. Carnevale F, Mendini M, Moirani G. Bernardino Ramazzini works. The diseases of workers; the health of princes. Verona: Cierra Edizioni; 2009.
2. Zanatta A, Zampieri F, Scattolin G, Rippa BM. Occupational markers and pathology of the castrato singer Gaspare Pacchierotti (1740-1821). Sci Rep. 2016;6:28463.

3. Loudon RG, Roberts RM. Singing and the dissemination of tuberculosis. Am Rev Respir Dis. 1968;98(2):297–300.
4. Hutcheon L, Hutcheon M. Opera: desire, disease, death. Lincoln-London: University of Nebraska Press; 1996.
5. De Cristoforis M. Annotazioni intorno la patologia e la terapia della voce specialmente considerata in rapporto all'arte del canto. Annali Universali di Medicina. 1867;63:321–51.
6. Poore GV. Clinical lecture on certain conditions of the hand and arm which interfere with the performance of professional acts, Especially Piano-Playing. Br Med J. 1887;1(1365):441–4.
7. Wahltuch A. Violinist's cramp treated successfully by electricity. Br Med J. 1886;1(1305):11–2.
8. Singer K, Lakond W. Diseases of the musical profession: a systematic presentation of their causes, symptoms and methods of treatment. New York: Greenberg; 1932.
9. Middlestadt SE, Fishbein M. Health and occupational correlates of perceived occupational stress in symphony orchestra musicians. J Occup Med. 1988;30(9):687–92.
10. Vaiano T, Guerrieri AC, Behlau M. Body pain in classical choral singers. Codas. 2013;25(4):303–9.
11. Cammarota G, Masala G, Cianci R, Palli D, Capaccio P, Schindler A, Cuoco L, Galli J, Ierardi E, Cannizzaro O, Caselli M, Dore MP, Bendinelli B, Gasbarrini G. Reflux symptoms in professional opera choristers. Gastroenterology. 2007;132(3):890–8.
12. Clarós P, Sobolewska AZ, Clarós-Pujol A, Pujol C, Clarós A. Prevalence, symptoms and treatment of vocal fatigue in professional opera singers: Clinician's experiences. Sur Res Rep. 2019;2(1):1–7.
13. Achey MA, He MZ, Akst LM. Vocal hygiene habits and vocal handicap among conservatory students of classical singing. J Voice. 2016;30(2):192–7.
14. Phyland DJ, Oates J, Greenwood KM. Self-reported voice problems among three groups of professional singers. J Voice. 1999;13(4):602–11.
15. Clarós P. Professional opera singers' voice disorders in relation to voice type. AMCRS. 2019;2(1):1–8.
16. Guss J, Sadoughi B, Benson B, Sulica L. Dysphonia in performers: toward a clinical definition of laryngology of the performing voice. J Voice. 2014;28(3):349–55.
17. Richter B, Löhle E, Knapp B, Weikert M, Schlömicher-Thier J, Verdolini K. Harmful substances on the opera stage: possible negative effects on singers' respiratory tracts. J Voice. 2002;16:72–80.
18. Schink T, Kreutz G, Busch V, Pigeot I, Ahrens W. Incidence and relative risk of hearing disorders in professional musicians. Occup Environ Med. 2014;71(7):472–6.
19. Zhao F, Manchaiah VK, French D, Price SM. Music exposure and hearing disorders: an overview. Int J Audiol. 2010;49:54–64.
20. Palin SL. Does classical music damage the hearing of musicians? A review of the literature. Occup Med (Lond). 1994;44:130–6.
21. Fligor B, Le Prell CG, Wartinger F, Portnuff C, Berger E, Santucci M. National Hearing Conservation Association (NHCA) position statement. Recreational music. Exposure. 2018; https://www.hearingconservation.org/assets/Decibel.pdf. Accessed March 1, 2022
22. Dinakaran TDRD, RejoyThadathil C. Awareness of musicians on ear protection and tinnitus: a preliminary study. Audiol Res. 2018;8(1):198.
23. Pouryaghoub G, Mehrdad R, Pourhosein S. Noise-induced hearing loss among professional musicians. J Occup Health. 2017;59:33–7.
24. Callahan AJ, Lass NJ, Foster LB, Poe JE, Steinberg EL, Duffe KA. Collegiate musicians' noise exposure and attitudes on hearing protection. Hear Rev. 2012;19:42–51.
25. Kahari K, Zachau G, Eklof M, Moller C. The influence of music and stress on musicians' hearing. J Sound Vibrat. 2004;277:627–31.
26. Di Stadio A, Dipietro L, Ricci G, Della Volpe A, Minni A, Greco A, de Vincentiis M, Ralli M. Hearing loss, tinnitus, Hyperacusis, and diplacusis in professional musicians: a systematic review. Int J Environ Res Public Health. 2018;15(10):2120.
27. Silva AG, Lã FM, Afreixo V. Pain prevalence in instrumental musicians: a systematic review. Med Probl Perform Art. 2015;30(1):8–19.

28. Knishkowy B, Lederman RJ. Instrumental musicians with upper extremity disorders: a follow-up study. Med Probl Perform Art. 1986;1:85–9.
29. Campos LGN, Pedrosa BH, Cavalcanti RVA, Stechman-Neto J, Gadotti IC, de Araujo CM, Taveira KVM. Prevalence of temporomandibular disorders in musicians: a systematic review and meta-analysis. J Oral Rehabil. 2021;48(5):632–42.
30. Kok LM, Groenewegen KA, Huisstede BMA, Nelissen RGHH. The high prevalence of playing-related musculoskeletal disorders (PRMDs) and its associated factors in amateur musicians playing in student orchestras: a crosssectional study. PLoS One. 2018;13(2):e0191772.
31. Nawrocka A, Mynarski W, Powerska-Didkowska A, Grabara M, Garbaciak W. Musculoskeletal pain among polish music school students. Med Probl Perform Art. 2014;29:64–9.
32. Ranelli S, Straker L, Smith A. Playing-related musculoskeletal problems in children learning instrumental music: the association between problem location and gender, age, and music exposure factors. Med Probl Perform Art. 2011;26:123–39.
33. Jacukowicz A. Psychosocial work aspects, stress and musculoskeletal pain among musicians. A systematic review in search of correlates and predictors of playing-related pain. Work J Prev Assess Rehabil. 2016;54(3):657–68.
34. Vervainioti A, Alexopoulos EC. Job-related stressors of classical instrumental musicians a systematic qualitative review. Med Probl Perform Art. 2015;30(4):197–202.
35. Yang N, Fufa DT, Wolff AL. A musician-centered approach to management of performance-related upper musculoskeletal injuries. J Hand Ther. 2021;34(2):208–16.
36. Shanoff C, Kang K, Guptill C, Thaut M. Playing-related injuries and posture among saxophonists. Med Probl Perform Art. 2019;34:215–21.
37. Yeung E, Chan W, Pan F, Sau P, Tsui M, Yu B, Zaza C. A survey of playing-related musculoskeletal problems among professional orchestral musicians in Hong Kong. Med Probl Perform Art. 1999;14:43–7.
38. Baadjou VAE, Verbunt JAMCF, van Eijsden-Besseling MDF, Huysmans SMD, Smeets RJEM. The musician as (in)active athlete? Exploring the association between physical activity and musculoskeletal complaints in music students. Med Probl Perform Art. 2015;30:231–7.
39. Kreutz G, Ginsberg J, Williamson A. Music students' health problems and health-promoting behaviours. Med Probl Perform Art. 2008;23:3–11.
40. Spahn C, Richter B, Altenmüller E. Musikermedizin: Diagnostik, Therapie und Prävention von musikerspezifischen Erkrankungen. Stuttgart: Schattauer Verlag; 2011. p. 153.
41. van Kemenade JF, van Son MJ, van Heesch NC. Performance anxiety among professional musicians in symphonic orchestras: a self-report study. Psychol Rep. 1995;77:555–62.
42. Spahn C, Echternach M, Zander MF, Voltmer E, Richter B. Music performance anxiety in opera singers. Logoped Phoniatr Vocol. 2010;35(4):175–82.
43. Fernholz I, Mumm J, Plag J, Noeres K, Rotter G, Willich SN, Ströhle A, Berghöfer A, Schmidt A. Performance anxiety in professional musicians: a systematic review on prevalence, risk factors and clinical treatment effects. Psychol Med. 2019;49(14):2287–306.
44. Galimberti A, Cena H, Campone L, Ferri E, Dell'Agli M, Sangiovanni E, Belingheri M, Riva MA, Casiraghi M, Labra M. Rethinking urban and food policies to improve citizens safety after COVID-19 pandemic. Front Nutr. 2020;7:569542.
45. Belingheri M, Paladino ME, Riva MA. Beyond the assistance: additional exposure situations to COVID-19 for healthcare workers. J Hosp Infect. 2020;105(2):353.
46. Belingheri M, Paladino ME, Riva MA. COVID-19: health prevention and control in non-healthcare settings. Occup Med (Lond). 2020;70(2):82–3.
47. Graf S, Engelmann L, Jeleff Wölfler O, Albrecht I, Schloderer M, Kramer A, Klankermayer L, Gebhardt F, Chaker AM, Spinner CD, Schwab R, Wollenberg B, Protzer U, Hoffmann D. Reopening the Bavarian state opera safely: hygiene strategies and incidence of COVID-19 in artistic staff during theater season 2020/2021. J Voice. 2021;S0892-1997(21):00393.
48. Hamner L, Dubbel P, Capron I, Ross A, Jordan A, Lee J, Lynn J, Ball A, Narwal S, Russell S, Patrick D, Leibrand H. High SARS-CoV-2 attack rate following exposure at a choir practice–Skagit County. Morb Mortal Wkly Rep. 2020;69(19):606–10.

49. Vance D, Shah P, Sataloff RT. COVID-19: impact on the musician and returning to singing; A Literature Review. J Voice. 2021;S0892-1997(21):00003–5.
50. Riva MA, Paladino ME, Paleari A, Belingheri M. Workplace COVID-19 vaccination, challenges and opportunities. Occup Med (Lond). 2021;kqab080

Part IV

ACT IV Finale

Representational Conventions and Varieties of Operatic Madness: The Gap between Dramaturgy and Diagnosis

10

Naomi Matsumoto

The present collection of essays explores the complex and multifaceted relation between opera and medicine from diverse angles: in Act 1, the ways in which illnesses and medical practitioners are represented in operatic works and the methods and effects of 'opera therapy'; in Act 2, the changing cultural, social and medical attitudes towards, and beliefs in, psychiatric and neurological anomalies behind their operatic representations and 'neuromusicologiocal' investigations into the effects of opera and operatic singing on the human brain; and in Act 3, the connection between mental illnesses and musical creativity and health problems of professional musicians. I first became interested in such opera-medicine interconnection when I began working on my doctoral thesis which was to trace the origins and early development of the 'operatic mad scene' [1]. Although most of the operatic works discussed in the present volume were written between the eighteenth and early twentieth centuries (e.g. Act 1, Chaps. 1 and 2), I found that there already was a flowering of mad scenes and insane characters in operas and musical dramas in the seventeenth century and the subject of insanity recurred with regularity throughout the history of that genre. This is somewhat contrary to our common belief that madness in opera came to the fore in the early nineteenth century—perhaps our idea is fixated too much upon famous instances from that period such as Imogen in *Il Pirata* and Elvira in *I Puritani* (both by Bellini), or Lucia in *Lucia di Lammermoor*, Linda in *Linda di Chamounix* and Anna in *Anna Bolena* (all by Donizetti). However, the tight connection between opera and insanity from its nascent stage is not particularly surprising if we consider that mad scenes and characters are ubiquitous in many ancient, medieval and renaissance literary and dramatic genres which directly influenced the formation of opera. For example, in Homer's *Illiad*, we find Achilles,

N. Matsumoto (✉)
Goldsmiths, University of London, London, UK
e-mail: n.matsumoto@gold.ac.uk

L. Lorusso et al. (eds.), *Effects of Opera Music from Brain to Body*, Neurocultural Health and Wellbeing,
https://doi.org/10.1007/978-3-031-34769-6_10

Agamemnon, Andromache, Bellerophon, Diomedes, Hector and Lycurgus and, in his *Odyssey*, the character Antinous, all of whom were driven to insanity. Greek tragedies such as *Ajax, Oedipus Tyrannus* (both by Sophocles) and the *Bacchae, Hercules furens, Iphigeneia in Tauris, Medea, Orestes* and *Troades* (all by Euripides) feature insanity in one way or another. In Roman literature, while Ovid's *Metamorphoses* recounts insane characters taken from the Greek myths, Virgil presents in his *Aeneid* Dido who is abandoned by Aeneas and driven to madness. Later in the history, the materials of the *Commedia dell'arte,* established in the sixteenth century, include several insane characters, usually employed for comedic purposes.

As a doctoral student I was initially hoping to discuss how contemporary medical practices affected and were reflected in the operatic representation of madness in the early modern period. However, I soon realised that my research plan was not quite working, not because of my rather limited knowledge about medicine, but because contemporary medical practices emerged very seldom in seventeenth-century operatic libretti. Also, the display of insane 'symptoms' in those works tends to be stylised and conventionalised rather than 'realistic'. They are conveyed not only by nonsensical utterances (which we may find 'typically and clearly mad') but, for example, endless references to Greek mythological gods and monsters (strongly reminiscent of the hallucinations that those mad characters suffer in Ovid's *Metamorphoses*), or by sudden and meaningless catalogue of a variety of things (not unrelated to the tradition behind Leporello's 'Catalogue aria' from Mozart's opera, *Don Giovanni* on the libretto by Lorenzo da Ponte). Other conventionalised tools to represent insanity include verse types employing rhythmic and metrical disruption (e.g. *sdrucciolo* [proparoxytone] endings and *novenari* [nine-syllable poetic lines], both frowned upon in traditional sophisticated Italian versification). Perhaps among relevant works within the seventeenth century, the only exception which contains some direct references to contemporary medical procedures is the anonymous 1-act opera *Lo spedale* (c. 1660) (for a modern edition and commentaries, see [2]). It presents a doctor who prescribes as traditional remedies for insane patients chickpeas or a laxative made of aloes, or even attempts to bash one of them with a stick. Although the patients initially praise him as the 'philosopher and astrologer in one' and equate him to Cicero and Galen, they condemn him when they realise he is motivated by a blatant greed for money. This chimes well with the examples discussed in Act I, Chap. 2, of this volume which explores stereotypical figures of medics in opera.

In the end, my doctoral work altered course so as to explore cultural and historical mechanisms that lay behind the fictive 'contract' between those insane characters and their audiences—in the seventeenth century and in our own times. However, what emerged is very much in line with the overarching theme of this volume: opera as a whole, in a symbolic manner, indicates our developmental understanding of behavioural anomalies as well as our attitudinal shifts towards those suffering from those conditions. First, 'mad' scenes from the early modern period present many similar but aetiologically distinct symptoms in a somewhat confused manner. What we find there is not only those whom doctors nowadays might diagnose as insane but also those suffering from melancholy and/or extreme lovesickness as well as intellectually challenged simpletons. This attests to an underdeveloped comprehension of those conditions in the contemporary world of early opera which also led to a belief in the mystical power that those with mental conditions thought to bear at that time. Moreover, that too-inclusive concept of the 'mad' in opera somewhat remained later in the nineteenth century when our neuro-psychiatric understanding improved. Opera kept categorising madness and other comparable conditions in a catch-all manner and, as a result, even nowadays, *The New Grove Dictionary of Opera* defines the mad scene rather uncritically as: 'an operatic scene in which a character, usually the soprano heroine displays traits of mental collapse, for example through amnesia, hallucination, irrational behaviour or sleepwalking' [3]. Aiding these simplifications is the fact that by the nineteenth century, opera had already established its representational conventions so that those conditions appear on the operatic stage with already settled representational tools. Thus, we need to be cautious in drawing easy parallels between the representation of the 'mad' in opera and wider but contemporary social views of madness, or with their professional theories of about the insane. Some such issues are already discussed in Act 2, Chap. 1, of this book.

That said, we can trace from the gradual changes which occurred in operatic representations some of the shifts in society's attitude towards the mentally afflicted. On the early modern operatic stage, all characters considered to be 'insane' are drawn so as to evoke laughter among the audience, rather than sympathy. Those protagonists often engage in antics together with other minor, comic characters even if their original characters before the onset of the condition were noble or regal. The attitudinal shift from objects of laughter and ridicule to those of sympathy and pity seems to have occurred first at the end of the seventeenth century when the focus of operatic madness moved from men to women. Initially in operatic works, male protagonists were often more susceptible to madness than their female counterparts. My research uncovered some 50 mad operas in the Italian language from the seventeenth century, and in 37 of them the mad characters are male. There seem to be two factors behind this. First, in Greek literature to which the formation of opera was indebted, many references to madness have a clear connotation of masculinity, because they very frequently appear in the context of battle scenes. Second, many operas written in the seventeenth century were loosely based upon Ludovico Ariosto's epic, *Orlando furioso* (needless to say featuring the insanity of the paladin, Orlando) published first in 1516, which thus gave rise to many Orlando and Orlando-like characters. In general, it was Roman literature where the first move towards female madness was noted. The most significant opera from the early modern period which shows clear indebtedness to Roman notion of female insanity is *Didone delirante*, premiered in Venice in 1696 (the music by Carlo Pallavicino on the libretto by Antonio Franceschi). This tragedy was derived from the Dido story by Virgil and, unlike preceding operatic examples, was designed primarily to evoke in the audience a sympathetic response. From then on, similar, piteous representations of female madness became common through the eighteenth and nineteenth centuries. Even so, certain 'psychological' differences in those representations are discernible, dependent upon the era. The madness of the title role in *Didone delirante* is portrayed particularly vividly in a scene where she unfolds a poignant dialogue with her own mirror image without recognising herself (Act III, Scene 17). That scene symbolises a notion of 'pre-modern' madness, which, as Gary Tomlinson has described, constitutes an 'externalization…nicely preserved in such Italian archaisms as "*fuor di sè*" [out of one's self] and "*forsennato*" [out of one's senses]' [4]. In other words, in the seventeenth century madness was some 'external state' where one might become estranged from one's usual self, whereas in later operas from the eighteenth century on, with its fragmentary psychological deconstruction of the personality in mad scenes, one is led to extreme forms of the interior chaos of the character revealing the fragile and illusionary notion of 'identity' rather than the outward behavioural aberration. Psychological deconstruction became even more apparent in nineteenth-century heroines such as *Lucia di Lammermoor*.

Moreover, in so far as the 'gendering' of insanity can be construed from the course of operatic history, it would seem to have arisen from contrasting attitudes towards males and females. The extremes of behaviours in men (including mad acts) in opera tend to be seen as temporary events, which sublimate rather than destroy their essential powers and the overall strengths of their personalities. Women, on the other hand, are seen as 'essentially' unstable, and their mad acts as symptoms of a mere general malaise and vulnerability. It is perspectives such as these that became the basis for Sigmund Freud's influential but notorious essay on the link between hysteria and women [5]. In musicology for some decades now, we have seen the methods and practices of feminism employed to form a major approach to the interpretation of 'mad' opera, in opposition to what is seen as the paternalistic Freudian bias [6, 7]. However, their assumption that there is clear tendency in the history of opera to treat only women as mad characters itself is a somewhat stereotypical bias—the reality is quite the opposite, as I have mentioned above.

Opera, with its diverse 'ingredients' (musical, literary, visual, dramatic, histrionic, visual, chorographical, etc) and many conventions established gradually over time, is a very complex artform as several chapters of the present volume reveal. Traditional musicology has long been aware of its shortfall in encapsulating the full extent of that genre and has encouraged interdisciplinary perspectives for opera. Given the resultant expansion of the methodologies during the most recent decade, we have seen a few attempts to review and 'codify' those methodologies for 'opera studies' [8, 9]. The methodologies consist, for example, of textual, musical, contextual, documentary, visual, dramaturgical and institutional analyses and examinations as well as hermeneutic approaches in relation to performance and performativity, gender, genre and nations and their specific cultures. Regrettably, however, despite the strong interrelation between opera and medicine, there is not yet a full rapprochement between the medical and musical disciplines. An obvious and immediate contribution of this book is its attempt to fill that void and offer those interested in opera perspectives given by medical historians and physicians from different branches—neurology, neurosurgery and occupational medicine. Such a collaboration between those in diverse fields is, perhaps, of particular importançe in these difficult times.

The pandemic with Covid-19 has been challenging to all of us in a variety of ways, but its impact on the musico-theatrical industry has been immeasurable, adding immense extra pressure on practitioners. Upon the implementation of lockdown or social distancing, opera houses across the globe had to devise measures to survive. Some sought a solution in temporary online/digital platforms (which might or might not acquire new audience members), while others restricted the number of audience seats, both facing the higher cost incurred and a drop in income. Some of those difficulties are discussed in Act 3, Chap. 3, in this volume. We can safeguard opera not only by keeping it on the stage but also by keeping its meanings and relevances alive by engaging in thought-provoking explorations of its cultural and societal significance from many viewpoints. Working with simple stereotypes may help us swallow the 'medicine' of the complexities of opera, but that is never going to help us understand the constantly evolving dialectic between artworks and the societies from which they emerge. Attention to detail will provide the path to discovery.

References

1. Matsumoto N. The operatic mad scene: its origins and early development up to c. 1700. PhD diss: University of London; 2005.
2. Matsumoto N, editor. Lo Spedale: Dramma burlesco. Middleton, WI: A-R editions; 2022.
3. Willier SA. Mad scene. Grove Music Online; 2002. Accessed 25 Jul 2022. https://www-oxfordmusiconline.com.gold.idm.oclc.org/grovemusic/view/10.1093/gmo/9781561592630.001.0001/omo-9781561592630-e-5000007756
4. Tomlinson G. Metaphysical song: an essay on opera. Princeton, NY: Princeton University Press; 2014.
5. Freud S. Studies on Hysteria. Translated by James Strachey. London: Vintage; 2001 [1895].
6. Clement C. Opera or the undoing of women. Translated by Betsy Wing. Minnesota: University of Minnesota Press; 1988 [1979].
7. McClary S. Feminine endings: music, gender, and sexuality. Minnesota and London: University of Minnesota Press; 1991.
8. Till N, editor. The Cambridge companion to opera studies. New York: Cambridge University Press; 2012.
9. Greenwald HM, editor. The Oxford handbook of opera. New York and Oxford: Oxford University Press; 2014.

SPRINGER NATURE

GPSR Compliance

The European Union's (EU) General Product Safety Regulation (GPSR) is a set of rules that requires consumer products to be safe and our obligations to ensure this.

If you have any concerns about our products, you can contact us on ProductSafety@springernature.com

In case Publisher is established outside the EU, the EU authorized representative is:

Springer Nature Customer Service Center GmbH
Europaplatz 3
69115 Heidelberg, Germany

The manufacturer's authorised representative in the EU is Springer Nature Customer Service Centre GmbH, Europaplatz 3, 69115 Heidelberg, Germany. If you have any concerns regarding our products, please contact ProductSafety@springernature.com

Printed and bound by CPI Group (UK) Ltd, Croydon, CR0 4YY

29/07/2026

02180513-0010